Bhopal Survivors Speak:
Emergent Voices from a People's Movement

Word Power Books,
43-45 West Nicolson Street,
Edinburgh, EH8 9DB.
Tel 0131 662 9112

www.word-power.co.uk

Designed by Leela Sooben.
Printed and bound in Scotland.

ISBN 13: 978-0-9549185-9-0
ISBN 10: 0-9549185-9-2

British Library Cataloguing in Publication Data.
A catalogue record for this book is available from the British Library.

Bhopal Survivors Speak:
Emergent Voices from a People's Movement

Bhopal Survivors' Movement Study 2009

Bhopal Survivors' Movement Study comprises
Eurig Scandrett, Suroopa Mukherjee, Dharmesh Shah,
Tarunima Sen and many named and unnamed Bhopal survivors
and activists who have contributed to this work.

Contents

Editor's Preface and Acknowledgements

Bhopal 1984 must remain in the memories of everyone who is old enough to remember it, and not only in India. The immense devastation and suffering caused through the instrumental logic of science and economics shocked the world. For myself, who had no connection with India, it changed the path which my life took, and I know that I am not alone. Twenty years later I encountered the survivors' movement, still campaigning for justice, and started doing some work in solidarity. The Bhopal Survivors' Movement Study, and this book, is the result.

1984 was the early days of the neoliberal globalisation which has dominated world policy for more than two decades, at least until the financial world came close to collapse. In December 1984, I was in the middle of a Botany PhD in Aberdeen University and active in a small way in the radical science movement. Science in the UK was a site of struggle in the politics of globalisation. Eighteen months after her second election win, Prime Minister Margaret Thatcher was in full thrust of her privatisation programme. A scientist herself, she proceeded to shift as much scientific research as possible onto business and force commercial motives into publicly funded research. In the midst of this battle, the horror of Bhopal illustrated the logical outcome of putting science in the hands of commerce.

After Bhopal it was clearly impossible to claim to be a neutral scientist. The choice was whether to be on the side of the inhuman pursuit of profit or else on the side of humanity. Five years later, I decided that for me it was too difficult to pursue an ethically justifiable scientific career. I left scientific research behind and became an adult educator. I worked in community groups, trade unions and environmental groups until, in 2004, whilst employed at Friends of the Earth Scotland, I went to the World Social Forum in Mumbai where I met activists in the International Campaign for Justice in Bhopal. Two years later, as an academic again, this time in social sciences, I got back in touch with ICJB, met Sathyu Sarangi and started to devise the research project which has led to this publication.

This book is the story of the struggle, told by some of those involved in it. As editor, I have tried to be faithful to the views of activists. As a result there are strongly expressed and often contradictory claims amongst the pieces. The research team has put great efforts into accurately reflecting the views of the contributors and if we have failed in that I must apologise in advance. We have certainly not attempted to adjudicate between claims, and none of the opinions expressed in this book should be regarded as being shared by anyone other than the person in whose name it is expressed.

I have been the outsider in this project, devising methodologies, raising funds, editing this book. The project has been produced by very many other people. The vast majority of the work has been carried out by two research assistants, Dharmesh Shah and Tarunima Sen, whose skill, humanity, enthusiasm and humour have provided the craftsmanship which produced the interview material. I am immensely grateful to them for their commitment to the project, as I am to Suroopa Mukherjee who was interviewing survivors well before I arrived in Bhopal. I am also very grateful to the survivors and activists, especially those in leadership positions who have taken the risk to invest so much trust in us with the stories of their lives. I hope that in their eyes this publication justifies that trust. I would especially mention Sathyu, Jabbar and Namdeo, without whose support none of the other interviews would have happened. Very many thanks to all those who have given us permission to use their stories in this book and to the many others whose confidential interviews have given us a deeper understanding of the issues. Anonymous versions of all interviews will be available in both Hindi transcription and English translation in public venues and on-line from Queen Margaret University archive http://edata.qmu.ac.uk/.

A special thanks to Maude Dorr, longstanding supporter of the Bhopal movement, who has contributed her outstanding photographic portraits for this book and to the great many people who have helped and supported the work, providing advice and space to conduct interviews, helping with contacts, translating between English and Hindi, proof reading and so on. Many thanks to all.

Funding for the research has come from the Nuffield Foundation, British Academy, Carnegie Trust for the Universities of Scotland, Barry Amiel and Norman Melburn Trust and Lipman-Miliband Trust to all of whom we are very grateful.

I would like to thank my colleagues at Queen Margaret University who have facilitated this work through their technical, administrative, fund-raising and academic skills, accommodated my absences in India and provided critical feedback at various stages of the work. And finally, I would not have been able to carry out my part in the research and production of this book without the support, encouragement and love of my partner, Susan, which is very much appreciated.

This book is a celebration of the achievements through struggle of a remarkable movement, but nowhere can the celebration forget the suffering on which it is based. Whilst researching the movement we encountered unbelievable strength, bravery and feistiness, but we also witnessed unimaginable pain, which 25 years has not made more bearable. Justice cannot be done in written words to the horrible deaths of wives, husbands, parents, children, loved ones, the theft of so many futures.

All we can do is add our voices to those of the survivors, activists and supporters throughout the world for justice to the people of Bhopal and a world where the events of December 3rd 1984 can never again occur. *Ladenge*, *Jeetenge*: we will fight, we will win!

Eurig Scandret
Scotland, July 2009

Glossary of terms and editorial notes

especially for readers from outside India

Hindi terms for protest movements and actions

Satyagraha nonviolence. More strictly 'seeking for truth' (*Satya* = truth; *Aagraha* = seeking). Gandhi's conception of a morally just way to fight injustice without resorting to collusion or violence in which Satyagraha constitutes the spiritual state of the activist and Ahimsa the political practice. The term Satyagraha is used for various protest actions which have developed using Gandhian resistance methods.

Dharna vigil, sit-in, encampment. Usually a long term occupation of space for the purposes of protest.

Padyatra long protest march. Literally, a long journey on foot.

Sangathan a group or organisation formed for the purposes of a campaign / protest

Morcha front, as in popular front, or a rally

Andolan movement

Sangh union

Ekta united

Gas Peedit gas affected

Pani Peedit (contaminated) water affected

Names of campaign groups

Hindi names of protest groups are given approximate translations in the text. These translations are often modifications of a strict translation in order for the sense to be kept in English eg *Bhopal Gas Peedit Mahila Udhyog Sangathan* is usually translated as Bhopal Gas Affected Women Workers' Union although a more strict translation might be Bhopal Gas Affected Women Industries Union.

Social work

The English term 'social worker' is used in India to describe a role which might range from community service volunteer or community worker to community organiser or social activist. It does not have the same meaning as a Social Worker in the UK.

Abbreviations commonly used

BGPMUS Bhopal Gas Peedit Mahila Udyog Sangathan [women worker's union]

BGPNPBSM Gas Peedit Nirashrit Pension Bhogi Sangharsh Morcha [Destitute pensioners' front]

BGPMSKS Bhopal Gas Peedit Mahila Stationery Karmchari Sangh [Stationery worker's union]

ZGKSM Zehreeli Gas Kand Sangharsh Morcha [poisoned gas event struggle front] the original movement organisation and regarded as the leadership for the whole struggle.

JSK Jan Swasthya Kendra [people's health clinic]

BGPSSS Bhopal Gas Peedit Sangharsh Sahayog Samiti [struggle solidarity committee]

ICJB International Campaign for Justice in Bhopal

NBA Narmada Bachao Andolan [Save the Narmada (river) Movement]

UCC Union Carbide Company. The parent company which owned the pesticide factory when the gas leak occurred.

UCIL Union Carbide India Limited. The subsidiary company which operated the pesticide factory. UCIL was established in India with controlling shareholding by UCC.

ICMR Indian Council for Medical Research

MIC Methyl Isocyanate

BPL Below Poverty Line

Other useful words

Lathi large wooden truncheon used by Indian police

Basti community, neighbourhood, an unorganised settlement

Mohalla a locality, in the context used here, a mohalla committee is a local area or neighbourhood committee.

Naxalite leftist guerrillas, usually Maoist, engaged in armed struggle in India. Named after armed peasant insurrection in Naxalbari, West

Bengal in 1967 led by a group which was to become Communist Party of India (Marxist-Leninist) [CPI(M-L)] and subsequently multiple factions with various names in different parts of India. The term Naxalite is more commonly used by state security forces and media to associate groups with extreme and non-state sanctioned violence, as in 'terrorist'.

Zamindar landowner in the Zamindari system of land ownership and taxation established during the Moghal period of Indian history, continued by British colonial regime and abolished in India at independence. Also Nawab: governor or nobleman in Moghal India; Jagir: plot of land granted by ruler.

Silai Sewing

Zari Fine gold or silver thread used in embroidery

Godown Warehouse, storage areas for a factory

Beedi Indian cigarette. Tobacco rolled in tendu leaf

Names and honorific titles for people

Bee honorific name adopted by married Muslim women
Bhai brother, often added to man's name out of respect
Appa older sister (Urdu), respectful name for older Muslim woman
Didi older sister (Hindi), respectful name for older Hindu woman

Indian numerical terms

Lakh hundred thousand 100,000 (UK) 1,00,000 (India)

Crore ten million 10,000,000 (UK) 1,00,00,000 (India)

Indian political parties

Congress Indian National Congress. Formed in 1885 and led opposition to Colonial rule from Britain, Congress was the party of Nehru and of uninterrupted government in India from 1947 to 1977. Secular, democratic and originally socialist, although now embraces neoliberal economics.

BJP Bharatiya Janata Party. Hindu Nationalist (Hindutva) party. Right wing, authoritarian and communalist, advocating intolerant and fundamentalist Hinduism in India, anti-Muslim, anti-Christian and anti-secular. Linked to various cultural, welfare and violent underground groups and networks. Governing party in Madhya Pradesh State.

CPI / CPM CPI is Communist Party of India, the older of the two communist parties in electoral politics in India. CPM is Communist Party of India (Marxist), which split from CPI in 1964, although the two parties often form alliances under Left Front.

Note on interview transcriptions

Most of the material in this book is derived from interviews with survivor activists, translated into English and then edited to form a narrative coherence. In some cases material is derived from more than one interview with the same individual, including a more analytical, follow-up interview which we called 'video diary'.

Three contributions were commissioned and written in English, and these are presented in a shaded box.

Introduction

Bhopal Survivors' Movement Study: Eurig Scandrett, Suroopa Mukherjee, Dharmesh Shah, Tarunima Sen

Bhopal 1984 must remain in the memories of everyone who is old enough to remember it, not only in India but also throughout the world. The leak of Methyl Isocyanate (MIC) gas from the Union Carbide pesticide factory in the early hours of 3rd December of that year remains the world's most devastating industrial disaster and incident of environmental pollution. It is a corporate crime of historic proportions. Estimates of the numbers of people who have died over the 25 years as a direct result of gas exposure are in the tens of thousands. What is surprising is that, amongst younger people, Bhopal is less well known. Moreover, that no company or individual has yet been prosecuted for the disaster is a travesty.

Throughout those 25 years, a remarkable social movement has been sustained, demanding basic rights from governments, tenaciously campaigning for justice from perpetrators and audaciously taking on multinational corporations and their logic of globalisation. The factory was a product of the 'green revolution' which gave corporations a foothold into Indian agriculture. Globally, 1984 was in the early days of the neoliberal experiment which dominated world policy for nearly three decades until the financial world came close to collapse. India's government came late to this global phenomenon and embraced inward investment from multinational capital from the 1990s. The Bhopal survivors' movement has been struggling against this tide.

Whilst the events in Bhopal have been the subject of a great deal of documentation and research, surprisingly little has been written on the experience of the survivors who have been actively campaigning for justice. Material has often been collected for the purposes of medical, legal or campaigning purposes and the gas affected population presented as victims. They are victims, but they are also survivors. People from the gas affected communities have mounted a sustained campaign for justice: for compensation, for economic rehabilitation, for adequate rations, for health care, for clean water, for environmental remediation of the factory site, for legal retribution against the company responsible and its directors. They have also established their own organisations to provide support and services, made resource demands on government and allies, conducted their own research and created their own traditions. Despite powerful enemies and many defeats they have celebrated their victories and their solidarity. The emphases, tactics and coalitions have changed over the years but there has been a consistent level of protest which is remarkable to anybody, and also fascinating for a researcher to study.

Most widely known of the campaign groups outside of India is the International Campaign for Justice in Bhopal (ICJB) which is a coalition of survivors' organisations, solidarity groups and campaigners across the world. It is not the only group campaigning for justice in Bhopal. Within India, the mass membership union *Bhopal Gas Peedit Mahila Udyog Sangathan* (Bhopal Gas Affected Women Workers' Union) has an impressive record of campaigning and service since its formation in 1986, at times in coalition with the smaller groups. The remaining survivors' campaign, *Nirashrit Pension Bhogi Sangharsh Morcha* (Destitute Pensioners' Struggle) predates the gas leak but since 1984 has focused attention on the survivors dependent on state pension. With the solidarity group, *Bhopal Gas Peedit Sangharsh Sahayog Samiti* (Bhopal Gas Affected Campaign Solidarity Committee), groups no longer functioning and activists formerly associated with these groups, this collection has attempted to reflect the breadth of the movement. The Bhopal campaign, like many mature social movements, is diverse and these groups often take different tactical approaches to, and adopt differing interpretations of events based on varying ideologies. It has always been the intention of this research to reflect this diversity without favouring one or other perspective and certainly without fuelling any disagreements. We hope that this publication reflects this objective, but any deviation from this representation is a result of our own shortcomings rather than a deliberate attempt to favour one position or argument.

Whilst we have meticulously remained neutral in the often fiery debates which take place within the movement, this is not neutral research. In the face of injustice on such a massive scale, it is impossible for research to be neutral. It has always been the contention of the research project that it is on the side of the survivors and hopefully will become part of the wider movement of solidarity with their struggle for justice.

Movement relevant research

So much academic research is irrelevant to the real world. It is designed to answer arcane questions of interest only to specialists in a narrow field and is published in journals which few people read. Even more research is carried out directly for commercial interests or to deliver state policy. It is so easy for this kind of research to abdicate responsibility for being critical. Research needs to be critical in two senses. First, it must be subject to the intellectual rigour of critical thinking – is it coherent and consistent, both within its own argument and in relation to the arguments of other academic writing? But secondly it must also be critical in the sense of an ethical questioning of its place in society. Whose interests does this research serve?

Because social research looks at society, and is itself part of society, it will have some kind of an impact on society. Neutral social research is not possible. Therefore, in order to have an ethical response to a social world which is structured according to injustices, a world which can allow thousands of the poorest in a poor country to be killed and maimed in a single night in pursuit of profits for shareholders in one of the richest countries, it is necessary to find a way in which that research contributes in some way to challenging these injustices. Not to do so merely perpetuates or even exacerbates it.

Academic research is often divided into pure and applied, or that which generates knowledge for its own sake, and that which generates knowledge for instrumental purposes. Whereas the legitimacy of pure research is measured only by its acceptance in the academic community, applied research is evaluated both by academics and by some kind of practical outcome – a bridge doesn't fall down, a disease is cured, a nation's economy grows. There is another type of research which might be called 'relevant'. Relevant research is evaluated primarily by a particular set of interests, a social and ethical choice about the impact of the research on society, asking the questions of who benefits and who loses. Now, applied and even pure research benefits some people whilst others lose out, but these kinds of questions are not a primary motivation for the research – indeed these kinds of questions are often not asked. When they are not asked, what tends to happen is that the benefit goes to those who already benefit in society – the rich, the powerful, the influential, the oppressing. For those of us who explicitly set out to do relevant research, the principal motivation is whether it is useful to the poor, the disempowered, the disenfranchised, the oppressed. Indeed, our aspiration is for research to be 'really useful' as distinct from merely 'useful', that is its usefulness must be judged by movements of the oppressed who are struggling for their own empowerment[1]. Thus the judgement of such research is not based on how many academic papers are produced in which journals, but how well the knowledge can be put to use by those who could benefit most from it.

A research team was put together comprising a range of skills. Eurig Scandrett is a sociology lecturer in Scotland, with a background in ecological research, adult education and environmentalism. Suroopa Mukherjee, an academic in the department of English, Hindu College, Delhi University and author of fiction and books for children, was at the time in receipt of a Fellowship at the prestigious Nehru Memorial Museum and Library, New Delhi, to conduct oral history with women survivors in Bhopal. Tarunima Sen was a Delhi based recent postgraduate in Sociology and Geography with an interest in the use of community arts for development. Dharmesh Shah was a full time environmental activist in corporate accountability and community environmental monitoring, based in Chennai. By the end of 2007, this multidisciplinary team started work collecting the stories of the survivors who have been active in the campaign for justice. All interviews were conducted by Dharmesh and Tarunima on the basis of ongoing discussions between the four of us.

Dharmesh, born just a year before the gas leak, describes what brought him into the project:

> "When I embarked upon the research, my understanding of Bhopal was metaphorical, often inspiring my environmental justice work in India. Bhopal was an impersonal concept that

I could borrow to argue my case for environmental justice and to question unsustainable development. As a campaigner with an environmental grassroots support organisation, my work mainly involved providing assistance to pollution impacted communities demanding clean environment, government transparency and corporate accountability.

Still sceptical about the impact of academic contributions towards social change, doing social research on Bhopal was an unlikely choice. This was also because of my perception about academics being dispassionate and subjective. The research altered my perception about academics and changed my relationship with Bhopal forever. For one the research hinged upon the Freirean problem posing approach that made the participants co-learners/educators in the process rather than just subjects. This was in tandem with my belief in community based knowledge/experience as being a rich resource. Secondly there was a commitment to the movement at the onset of the study in terms of providing regular feedbacks on the progress and offering complete publicity rights of the interview to the participants.

The research was put into perspective largely by the initial introductory sessions on social theory. This proved extremely beneficial as a foundation exercise that introduced us to concepts and techniques. A key research technique was based on the Freirean problem posing concept. This gave me a great deal of freedom as a researcher in the field. I could source techniques from my experience in community work and apply theories I learnt here to my work.

Social research which tries to get inside the culture and social relationships within a group of people is called ethnography. The classical tools of ethnography are the interview and the participant observation. Participant observation means that the researcher joins in with the social relationships which are going on whilst at the same time reflecting on them in order to learn more about them. This insider-outsider position of the researcher allows them to use their skills and resources as a human being as well as a trained social scientist to glean lessons about how the group works. Interviews are a slightly more systematic method of finding things out about the groups. They are usually 'semi-structured', which enables the researcher to guide the discussion but also allows for the interviewee to talk about other things which seem relevant to them. Of course it is rarely possible to be a complete insider / outsider in a group since the presence of a researcher, no matter how familiar they are, will change the social relationships which are being studied, and a great deal of discussion goes on amongst sociologists and anthropologists about the significance of this.

In the early discussions amongst the research team we decided to build on these ideas from classical ethnography. In one sense, each of us was already a participant observer in the wider movement for justice for Bhopal. We had all, between us, attended demonstrations and rallies, organised protests and solidarity actions, supported activists, written books and popular articles and given speeches about Bhopal in different parts of the world. Our sympathy with the cause was presumed. During the research we all participated in rallies and dharnas organised by the various campaign groups, not simply in order to research them but because we wanted to support their demands. Educated outsiders, including white Europeans and North Americans as well as Indians, participate in some of these protests on a regular basis and we were extending this. For some of the more local protests which were not made public beforehand, we posed as journalists or passing tourists unexpectedly drawn to the spectacle. So we extended our participation into the local movements and especially so for Dharmesh and Tarunima who for ten months were absorbed full time in participation. At the same time we were constant observers, reflecting what we were learning from the movement and from our studies, back into attempts to interpret what we were participating in. These observations were recorded as field notes, a discipline which Dharmesh and Tarunima developed and kept in personal journals, as well as in emails to Suroopa and Eurig. These field notes also provided a source for Dharmesh's and Tarunima's inputs here.

Secondly, the classical ethnographic method of semi-structured interview was adapted in order that the survivor-activists would be contributors to the research and not merely subjects or sources of information. Movement activists develop critical analytical skills through their practice and through reflecting together, collectively on that practice. This is true whether the activists are formally educated or, as is the case for many in the Bhopal movement, not literate. Thus in our attempts to gather material for this research, we developed a method based on interviewing and re-interviewing movement activists, using video recordings of their interviews as a tool for pursuing more critical lines of enquiry. We wanted to move a little away from an interview and more towards a dialogue between researcher and survivor. This approach draws on the educational methods of Paulo Freire, whose innovative approach was designed to value and build on the knowledge and interests of non-literate people in literacy campaigns.

Paulo Freire was a Brazilian educator who developed an approach to education which put the curriculum - the knowledge which is included in education – at the service of those who are struggling for political liberation. His *Pedagogy of the Oppressed* has been a valuable tool for many adult educators who see their craft as a tool in working for social justice. His methodology has also been influential for researchers looking for ways to do relevant research. As an adult educator, Eurig had used Freire's methods in various ways for many years, and thought they would be useful in the research context in Bhopal, where the objective is to produce knowledge which can be judged and used by a movement of people, most of whom have had very little formal education and are unable to read or write their spoken language.

Freire argues that people live in what he calls a thematic universe, an interconnected weaving of knowledge, experience, meaning, symbols and priorities. The main context for his work was teaching illiterate and landless Brazilian peasants to read and write in their mother tongue of Portuguese. But the thematic universe of a secure, middle class, educated teacher is so different from that of the peasants, that teaching them to read and write whilst ignoring their context risks treating people as though they were ignorant. The peasants are not ignorant. They cannot read the language which they speak. They are largely ignorant of the teacher's thematic universe, just as the teacher is ignorant of the peasants' thematic universe. The teacher does not know how to make crops grow in marginal land, or the experience of being humiliated by the landowner's security men at the funeral of your own child. So the first step for a teacher is to study the thematic universe of those who are being dehumanised by their social context and with them, find ways to analyse this world with a view to changing it for the better. In the process, both teacher and peasant become more fully human.

Tarunima describes what happens when theory interfaces with practice:

"Being a Sociology student I had heard and read a lot about movements and I was sure of what I knew through their definitions and references. But what lay ahead in my months in Bhopal was very different from the theoretical expression. How you see a movement through books is very different from what unfolds right in front of your eyes: the theory makes sense and answers many questions when you see it in practice. You read about the traditional leaders and organic intellectuals, their different ways of working and conflicts, and this is evident in many ways in Bhopal. On one hand there was a *sangathan* which spoke of local power and local supporters. And on another you had those who had built a campaign and were taking it forward with international support. I could also see the relevance of Gramsci, in the way that the role of power and knowledge changed in the course of development of the movement over the years. With more knowledge and information from both within and outside the country, the knowledge of the survivors became a source of their own power and of opposition to the structural power of the multi nationals and government."

So this is the objective of our research – to identify themes from within the thematic universe of the survivor-activists of Bhopal, and to present these themes back to them, with the addition of potentially useful knowledge from outside, in a way which they may be able to put it to use in their struggle for humanity.

The first stage of the method involved starting to build an informal picture of the thematic universe of the survivors' movement through Dharmesh and Tarunima making contact with the campaign leaders, attending public meetings and spending time with activists. This was an important trust-building process, especially given the degree of suspicion which some people in the campaign groups view the others. Sometimes this informal process yielded information that wouldn't emerge during an interview. It also helped to shape the questions which would be used with the interview.

Dharmesh: In the context of Bhopal it was important to explain to the survivors, the benefits of an oral history recording exercise like ours, primarily in sociological terms. Among the participants there was a sense of scepticism towards interviewers who they had encountered previously, most of whom were journalists or students they had interacted with but never saw to return after reporting. Hence, building trust among participants was perhaps the most important part of the study.

The most interesting challenge here was to explain the need for/importance of social research on the movement. Since each campaign group had a unique area of work we had first to get a thorough understanding of the history of each group through newspaper and document archives. The next step was to explain the research in the context of each group's frame of work using the right language.

Historically, the tragedy is a landmark and the survivors are its historians. This became the crux of our presentation to the various groups whose participation was key for the success of the project. The next step was establishing contacts preferably through neutral leads.

And typical of any social movement, ideological differences among the various campaign groups were prevalent. In such a situation while it was mandatory to establish neutrality between the differences, documenting this diversity was also important. In our situation the team's close association with the Sambhavna Trust, an organisation often associated with the International Campaign for Justice in Bhopal, made this process more challenging.

Through a mixture of careful neutrality, preparatory work, patiently passing tests of loyalty, referring where necessary to the authority of the two academic institutions to which we were affiliated (Queen Margaret University and Nehru Memorial Museum and Library), and an enormous amount of charm, Dharmesh and Tarunima managed the considerable achievement of being trusted by all groups sufficiently for activists to participate in the project and consent to being interviewed. They were able to get people to make the connections without animosity or sense of being sidelined, and to do this with competence and compassion. They reached the point where they were welcomed wherever they went and they won the confidence of people by encouraging them to speak out without fear.

Then when the interview was to take place the camera was set up, with additional microphones, sometimes in the premises of the campaign group or associated organisations, but more often in people's homes. The semi-structured interviews had been developed to bring out particular issues which the research team thought might expose interesting themes on the basis of prior observation, the academic literature and insights from longstanding activists: first involvement in the movement; what should be prioritised; how decisions are made and communicated; what events most memorable; what changes have occurred; why women are majority in movement; impact on religious practice; opinions about development in India; what would constitute justice etc.

These interviews often used triggers as memory stimuli, sometimes used in the way of what Freire called 'codes'. A stimulus could be any artefact or prompt which encourages people to remember things which they might otherwise not have remembered. This might be a photograph, a newspaper headline, a banner, or simply an event as recorded on a 'people's calendar' (a calendar of events which are widely remembered and which might be used to date other events which are less well remembered). A code is more than a memory stimulant; it is a simple representation of a familiar situation which, through asking questions about it can reveal deeper meanings and patterns of social reality. So for example, with some activists, we asked them to show us any photographs they had from the struggle, with others the researchers provided newspaper cuttings with strong visual images. The codes themselves stimulate stories about the people or events captured in them, but through sensitive questioning can also reveal more critical interpretation and analysis of the situation. The interrogation of the code tends to move from descriptive questions (what is there? who is that?) to more interpretive questions (what did you feel at the time? what happened next?) to analytical questions (why do you think that happened? how would you do it differently if you were doing it now?).

Thus a video was captured of a survivor-activist talking about her experiences of the movement, through a combination of the use of codes and prepared semi-structured interview questions, tailored where appropriate to issues which had arisen in earlier conversations. The default condition of the interview was that this video would only be seen by the research team and the interviewee, although some activists, including all of the interviewees in this collection, were happy for their interviews to be made public. Once videoed, a summary of the main points from the interview was made in English, shared amongst the research team who would discuss it and identify interesting issues arising. As the number of interviews accumulated, the discussions also addressed themes emerging from the whole research process, including the field notes of Dharmesh and Tarunima.

These discussions of individual interviews and emergent themes fed back into the interview process. The videoed interview was burned onto disk and a copy provided back to the interviewee so that they could view it at their leisure. At a later date, a secondary interview was conducted with many of the interviewees, the questions being based on the first interview and the themes which were emerging. Sometimes this interview took place whilst viewing the recording of the first interview and asking follow-up questions and digging deeper into the meanings of certain things. (This second interview became known as a 'video diary' because we believed that it could eventually take place without the presence of the researchers and become a self-generated diary controlled by the survivor-activist. In fact this never materialised within the timeframe of the research but the term 'video diary' stuck.) Thus, through a dialogue between the researchers and the survivor-activists, a progressively more interpretive and analytical picture of the movement emerged, and data accumulated on digital video. The next step in the process was to transcribe the interviews, translate them into English and edit them to remove means of identification so that they could become public data[2].

As Dharmesh and Tarunima became more familiar with using these techniques, they were internalised into their interactions with the survivors and so method became an extension of practice. Triggers or memory stimuli became 'probes' as the interviewers developed dexterity in using them to obtain more analytical reflections on events and, as Dharmesh describes, were later dispensed with as the interview content itself provided the opportunity for searching questions:

> **Dharmesh**: The praxis of the study was to initiate the process of reflection through a series of interviews and video diaries. These primarily emerged through an evolving set of probes or questions that facilitated the process. Initially we also relied heavily on these probes as a guide through the interviews but soon shifted our dependence on cues that emerged from the interviews.

As this process developed it acquired some variation. For some people the presence of the video camera was a deterrent to speaking freely and in certain other circumstances it was not possible to use the camera, in which case voice recorders were used to capture interviews. This was particularly so amongst the rank and file activists, less used to speaking in front of a camera. The progress of the interviews for each campaign group started with the leadership and mobilising cadres and then worked down through the ranks towards the rank and file membership and supporters. It is rare that campaign groups are structured quite as clearly as this but as a general trend, that was how contacts were made and interviews conducted. The rank and file, generally, were less well informed about overall movement strategies and historical events but their perspective on their involvement was essential. They were also more difficult to find and interviews were better conducted on the hoof wherever they could be identified, in public meetings or at rallies. However, interviews were collected from rank and file activists and contributed to the overall data to be transcribed and translated.

Some of the challenges of conducting interviews are described by Dharmesh:

At the stage of the interview the most important challenge was the use of technology. Since many of the participants were not used to interviews on camera, its presence caused great intimidation. The solution in such situations was to make interviews into conversations, sometimes even include small talk as a tool to eliminate the tension. As we progressed techniques had to be modified, in some cases we decided to spend several days accompanying/visiting (camera shy) participants in order to assess their temperament and later interview them.

Soon we also introduced pre-talks into the process, which were short/long conversations with participants before the video interviews: this was as an additional information gathering technique. Pre-talks were often audio recorded and later summarised for comments and suggestions from the team members on the issues to dwell on during the actual interview. Since audio recording proved to be less intrusive this was a good way to capture certain participants but less useful from a research perspective in which we proposed to use the videos as a tool for facilitating the process of reflection through video diaries.

Finally, some very basic survey data were collected from amongst attendees at public meetings and rallies, including such information as religion, marital status, occupation, caste and regularity of, and motivation for, participation.

In addition to these one-to-one interviews with survivor-activists (or on occasion with two survivor-activists together), after about 8 months into the project a number of group events took place. These were motivated by several factors. We thought that in some contexts a group discussion would generate information that individual interviews would not. Secondly, where new initiatives started during the period of our research and a group opportunity presented itself as a means of recording history as it was being made. And most significantly, we were always committed to being accountable to the movement and sought opportunities to feed back our reflections and insights, and where appropriate present issues from the academic literature on social movements. By the time these were carried out themes were already emerging which we wanted to explore in the context of discussions amongst survivor-activists. The group discussions were as follows:

Focus group interview with 7 mobilisers from ICJB (after 8 months), particularly focusing on how decisions are made and how knowledge is communicated and translated throughout the various parts of the campaign.

Symposium with 25 survivor-activists from all campaign groups (after 8 months), incorporating an opportunity for the research team to feed back some of the emergent themes from the research via English – Hindi translation by Dharmesh, and using pictoral representations of themes, courtesy of Tarunima's artistic skills.

Group interview with 8 members of Children Against Dow Carbide (after 17 months), a new campaign which emerged within ICJB whilst the research was underway.

Women's workshop with five women with influential positions in three campaign groups (after 17 months) in which these key women were invited to reflect on and discuss some issues from the research and some insights from social movement theory which may be of use to the survivors' movement.

The video-interview-dialogue process formed the core of the data collection and has amassed over 50 hours of interviews, whilst basic data from rank and file have been gathered for 119 individuals. Film footage has also been collected from rallies, protests, public meetings and dharnas which can be used to analyse participation. These, and the discussions amongst the research team, field notes and personal reflections constitute a significant amount of data. Add to this the oral history interviews conducted by Suroopa for her own research and this study constitutes a crucial contribution to the record of this unique social movement and will be a support to the movement in its campaign for justice. Social research has been described as like bricolage[3], or quilt making, in which multiple versions of events and meanings of phenomena are woven together to produce something of aesthetic, functional and interpretive value.

We hope that this will be the legacy of the research. But more importantly we hope that the process which we have followed by using our methods has been of value to the movement. The survivor-activists have contributed to the reflexive process which they bring to their campaigning and, we trust, have learned from their own experience as reflected back to them through ours.

If we succeed in this, even partially, then it will have been a valuable exercise. However we are ambitious for more. We believe that the insights which can be gleaned from the Bhopal survivors' movement will also yield lessons for other movements, and other researchers and activists who are supporting those movements. Social change for justice will only come when there is a critical mass of movement activity, all making claims to an economic and political system which is failing to deliver, indeed is systematically denying justice. Any study of an individual movement and the thematic universe in which it exists, must take cogniscence of the wider processes of economic change which the world capitalist system is undergoing.

The victims of Union Carbide / Dow in Bhopal are not simply oppressed by the gas leak and its aftermath. That is the mechanism of their oppression, not the cause, and remains the world's most devastating example of its type. However it is part of a family of mechanisms of oppression which have common cause in the logic of economics, which is that economic development will always try to maximise its gains. Wherever benefits outweigh costs, as measured in financial value, there economic development can occur. Such logic always requires costs to be shifted in as cheap a way as possible. What this means is that the locating of polluting and damaging industries, the cutting of corners, the means of neutralising resistance all lead to additional costs being borne by the poorest who have least leverage in an economic system driven by finance. Whether poisoning by toxins, working conditions, land dispossession, occupation of space or neglect of infrastructure, the experience world over is that it is the poor who suffer. But these poor also resist, and movements like Bhopal constitute just one part of a big picture of resistance to the economic logic of unfettered capital expansion. By documenting, reflecting back and contributing to the Bhopalis' struggle, we hope to contribute in some small way to the wider, worldwide struggle for justice, through analysing and interpreting movement activity in such a way that it becomes more successful in putting limitations and constraints on the destructive activities of capital and ultimately undermine its logic.

Timeline of the movement

A timeline is vital in any study of the aftermath of an industrial disaster, but even such a seemingly straightforward idea as a chronology of events is contested. Our project was able to draw on certain categories for classifying events, which were based on new directions taken by the social movement, as well as alterations in State policy, the role of international stakeholders and the changing global economic scene. This book has been conceptualised keeping in mind a number of thematic links that are directly drawn from individual case studies and emergent from across the movement. The approach is to generate and disseminate 'really useful' knowledge that serves the instrumental purpose of offering an alternative time scheme that emerges from within the voices of people engaged in the struggle.

Our interviews were conducted against the backdrop of the approaching 25th anniversary which became one of the important 'memory stimuli' that helped to create an episodic sense of events that could counter state level denial. Right from the start an official timeline was maintained by the establishment in order to keep a record of what kind of economic and medical schemes were started by Madhya Pradesh and central government, and the period of time before such schemes were closed down. This official timeline was contested by the actions of survivors which created a reservoir of their own memories that are both linear and overlapping[4]. There have been repeated attempts to dismiss people's complaints as false or exaggerated. Thus the various survivor organisations that have been identified here made a conscious effort to document their activities by keeping a record of newspaper clippings and photographs. We used these resources to trigger off the exchange between the interviewer and the interviewee. What we have tried to do in this collection is focus on the learning graph that becomes the marker for people's timeline.

The first phase following the disaster stretches from 1984 to 1989[5]. The official timeline identifies some important events: 'Operation Faith' on 16th December 1984, the much publicised event to "make safe" the MIC that remained in the plant; Government of India passed the Bhopal Gas leak Disaster (Processing of Claims Act) on 29th March, 1985, which gave the government the right to fight the legal case on behalf of thousands of poor and illiterate victims; early 1985 the Indian Council of Medical Research (ICMR) undertook studies to determine the long-term effects of the gas on the human body, keeping the findings confidential; 5th September 1986 the litigation against Union Carbide was transferred from the New York District Court to District Court in Bhopal, on grounds of "inconvenient forum"; at the end of 4 years of bitter legal battle the curtains were finally drawn with a Supreme Court ordered settlement for $ 470 million to be paid as compensation money to the survivors[6].

Soon after the disaster the government of Madhya Pradesh set up 50 training and production centres to train gas-affected women in a variety of trades such as sewing, embroidery and stationery manufacturing. In 1987, a special industrial area for training and employment of over 10,000 survivors was inaugurated and 152 work sheds were constructed at a cost of Rs. 8 *crore*. Construction of the sheds was completed in 1991. By 1992 these programmes were terminated and over 55 of these work-sheds were handed over to private enterprises and NGOs[7].

Post settlement the different phases become more amorphous and are marked by landmark events that do not merit prominence in the official timeline. Shortly after the settlement was announced, a general election brought an end to Rajiv Gandhi's Congress rule and the newly formed National Front government under Janata Dal's VP Singh acted to disburse the first tranche of compensation on the basis of Rs. 200 per month per person in the affected wards. From 1990 to 1995 People's Courts were set up to assess compensation claims and disburse the money to individual claimants. In October 1991, the Supreme Court passed a revised order on the settlement issue, by reinstating criminal cases against all the 12 accused. In April 1992 Warren Anderson, the Chief Executive Officer of Union Carbide Corporation (UCC) at the time of the disaster, was declared an 'absconder' and fugitive from law. In 1994 ICMR officially wound up all the projects without making the findings public. The same year the Supreme Court allowed UCC to sell off its encumbered assets in India and build a memorial hospital in Bhopal with the proceeds. In the absence of assets in India it became increasingly difficult to enforce criminal proceedings against UCC. In 1999 Greenpeace tested the soil and groundwater in and around the derelict factory site and discovered 12 toxic chemicals and mercury at levels that were 6 million times higher than the accepted level.[8]

Nearly two decades after the night of the disaster, in the year 2001, the Michigan based company Dow Chemicals, bought UCC for $ 11.6 billion and became the second largest chemical company in the world. While Dow accepted Carbide's liability in the Texas asbestos case originally filed against Carbide, it refused to do the same in Bhopal on grounds that it was not responsible for a factory it did not operate. In October 2002 the Madhya Pradesh government announced its decision to petition the Supreme Court to compel Dow to clean up the contaminated soil and groundwater at the factory site. This was based on a number of scientific assessments made by government and non-government organisations subsequent to the Greenpeace Report, confirming the depth and spread of toxic contamination. A new category of gas victims suffering from the consequences of water contamination had to be recognised by the government. In 2002 Government of India started the process of filing an application for the extradition of Anderson from the US. In 2004, in time for the official recognition of the 20th anniversary of the disaster, ICMR came out with its Epidemiological

Report that made a tame attempt to finalise the health effects of the toxic gas leak on thousands of survivors[9]. In 2004 the Government also announced with a lot of fanfare, the disbursement of the second and final instalment of the compensation money from the balance of $300 million, which had accrued from the appreciation and interest on money that had not been paid.

In a sense the official timeline has posed an embarrassment to the government, for it is proof enough of the continuity of the disaster and the lack of political will on its part to reach for long-term solutions to the problems. It also speaks of procrastination, failure to implement schemes and implicit support of the corporate logic that places profit before people. It is in this context that the people's timeline acquires significance. It creates necessary 'ruptures' in the official discourse and puts the onus on the government to come clean before people in order to keep the mechanisms of electoral politics in good health. The people's timeline points to what is missing or distorted in the government claims. It also shows how survivors were compelled to counter such claims with the creation of their own knowledge base.

We have tried to shape this collection by drawing on this knowledge base and showing how it has evolved over the last 25 years. There are a few themes we have identified, which run as a leitmotif through the interviews. The attempt has been to give even space to each of the themes, keeping in mind the constituency of speakers and the range of issues they talk about. It is important to remember that interviewees give both a personal and collective perspective, and despite the fact that they were videotaped, they agreed to the parameters of confidentiality and spoke candidly. Group leaders have contributed essays, some of which have been written in English as commissioned, whereas others are derived from translations of Hindi interviews and agreed through negotiation. Our purpose has been to capture the dynamics of the movement through memory trace, self analysis, arguments and counter arguments, dialogue and an exchange of ideas at the local/micro and global/macro level.

The people's timeline is more cyclical than linear; only one event is marked on the calendar every year, 2nd - 3rd December. There are some events – both onetime and recurrent – that come back to haunt the survivors, such as the 1989 settlement, the 2001 merger between UCC and Dow, the repeated threat to dislocate people from the *bastis*, the discovery that drinking water is unsafe, and the constant refusal on the part of the company to take on the onus of clean up. These are much debated issues and the survivors live with the constant fear of being victimised over and over again.

The timeline also works as a yard stick to understand the survivor's personal growth, his/her involvement with the organisations with which he/she is affiliated and the role they have played in shaping the movement. Here again dates have some importance but the people's calendar was far more instinctive and soon became a vital part of the interview method we used. In our interviews we have several examples of remembered dates which are factually incorrect but reveal a more significant truth of what was going on behind the scenes. An official announcement might be the final point of years of rumour or preparatory activism, or may be the stimulus for years of protest activity, so the date of the announcement becomes partially irrelevant to the people's timeline.

The starting point of the people's timeline tallies with people's memories that went back to the morning after the disaster. Immediately following the disaster there was a spontaneous outbreak of angry, unfocused protest amongst those who had survived, in amongst the frantic hunting for lost loved ones or panic about the sick. Within days, action was driven by necessity and became more focused, marked by *basti* level organisation of groups as well as voluntary help that poured in from all over India. Organisation focused on delivery of basic needs of food, water, shelter, comfort. Once the outsiders went away, groups of grassroots workers and intellectuals and activists with long-term commitment, were left behind to carry on with the work. They came together under different banners that represented different ideological and political affiliations: liberal and religious through Gandhian, socialist, communist and ultra-left. Most organised into the *Zehreeli Gas Kand Sangharsh*

Morcha (Poisonous Gas Incident Struggle Front), whilst those with affiliations to the Communist Party formed *Nagarik Rahat aur Punarvaas Samiti* (Rehabilitation and Welfare Committee). Activism, with its brand of conflicts and support building, began to charter demands based on contingency factors; the need of the hour was medical relief and taking care of the issue of livelihood. In these early days a relatively united movement could raise demands and seek concessions from government, and attempted to fill the gap in government services with the People's Health Clinic. Militant confrontation was met by government accusations of naxalism and subsequent violent repression. Clearly activism took different directions as the survivors' needs changed.

With the establishment of the government rehabilitation schemes in 1985, aimed largely at women, the next phase of activism focused on workplace organisation as wages, terms and conditions, corrupt practices and ultimately the closure of the work sheds were challenged. Women with no history of industrial action or even work outside the home, learned the disciplines of trades unionism. Unions such as the small but longstanding *Bhopal Gas Peedit Mahila Stationery Karmchari Sangh*, the short lived Communist affiliated *Gas Peedit Mahila Udyog Karmchari Ekta Parishad* and the mass mobilising and sustained *Bhopal Gas Peedit Mahila Udyog Sangathan*.

Activism divided itself into protesting government rehabilitation schemes at the local level and a legal battle against the Indian subsidiary and the US multinational. It is convenient to use the decade as a marker to indicate how needs and demands changed over the years. 1984 – 1994 saw important changes with the 1989 settlement as an important indicator of how the government was taking the side of the powerful corporation. The maximum outrage followed the settlement which was negotiated outside the courts but announced through the Supreme Court and regarded universally as a sell out, even though groups disputed the appropriate tactic to adopt. Since then groups have undergone schism around the tension between the grassroots nature of organisations and the need to get international support for the larger cause of human rights violation. Given the nature of a movement, which is dependent on educated intellectuals largely from the middle class, who rely on mass grassroots support, the movement had to walk a tightrope that often saw widespread disagreements, conflict of interest and change in the constituency of organisations. Our interviews have caught the dynamics of such change.

1995 – 2005 saw India's neo-liberal policies having its impact on the movement. The knowledge graph of the poor, often illiterate, survivors underwent a mammoth change. Some of the smaller groups united with international supporters into the International Campaign for Justice in Bhopal (ICJB) and focused on the new environmental front whilst *Bhopal Gas Peedit Mahila Udyog Sangathan* (Women Workers' Union) consolidated its grassroots focus on compensation and economic rehabilitation. Activism for some took on different tactics, using Gandhian methods of hunger strike, *padyatra* (long march) and peaceful processions through the main roads of Bhopal, at Jantar Mantar in New Delhi and in front of the corporate office in Mumbai of new owners of UCC, Dow Chemicals. The idea was to alert the world to the possibilities of more Bhopals happening in their backyards. At the same time, *Bhopal Gas Peedit Neerashrit Pension Bhogi Sangharsh Morcha* (Destitute Pensioners' Struggle) retained its focus on the most vulnerable groups which are entirely dependent on the state: old age pensioners, widows, the severely disabled.

The inclusion of the water contamination issue was a crucial move, for it made it very clear that the 1989 settlement had not laid the ghost of Bhopal to rest, and both company and government had many more tasks to fulfil. ICJB focussed on ongoing pollution, water toxicity and remediation of the contaminated land, Greenpeace got involved and this group became increasingly visible internationally as an environmental justice movement, at times shadowing other local groups. In 2004, Rasheeda Bee and Champa Devi Shukla, leaders of the *Bhopal Gas Peedit Mahila Stationery Karmchari Sangh* (Stationery Workers' Union) and part of ICJB were awarded the Goldman prize for environmental campaigning, an event which was both decisive and divisive. On the one hand, poor uneducated, gas affected women could win a prestigious environmental award, and with the money established Chingari Trust for work with disabled children and to support other struggles. On the other hand the decision split the Stationery Workers' Union and caused resentment from other, equally poor, uneducated and gas affected women active in groups which had taken a principled stand against international funding. Further tactical disputes emerged around whether Dow should clean up or pay for cleaning up the factory site, or whether, in the interests of speed, the state government should do so.

2005 and still continuing – has brought to the forefront the most serious consequences of an industrial disaster. The survivors are now facing the reality of another generation being born with ill health and congenital defects, and environmental degradation whose reach and spread is still unknown. The health issue is paramount and recently ICMR has admitted that there is a need to reopen some of the projects. Bhopal is supported by worldwide environmental groups and the issue of corporate liability has taken centre stage. An older generation of activists are now creating space for youth involvement. There is a do or die spirit that suggests the movement will not fizzle out with the passing generation; our interviews try to capture this hope that accompanies failure and victory.

The movement we encountered was squabbling and struggling but vibrant and creative. The differences amongst groups reflected the niche demographies of their membership, the strategic approaches of leaders and the discourses with which they interacted with others. For example, our data suggest that supporters of the different groups are drawn from different demographic groups. Whilst all groups are supported primarily by women of both Muslim and Hindu communities (and occasionally Christians), their supporters differ in their class, caste and status as gas- or contaminated water- affected. At risk of over simplifying, BGPMUS (Women Workers' Union) comprises gas affected working age Muslims and caste Hindus especially from Other Backward Classes (OBC); BGPNPBSM (Pensions movement) comprises older women, particularly from scheduled castes (dalits), and ICJB garners support from communities affected by water contamination across a range of castes.

Movement leaders articulated to us the purposes and tactics of meeting the interests of survivors in ways that reflect what might be called different ideological approaches to community development and social action. Take for example the following, admittedly simplified pencil sketches of the following campaign leaders.

Alok Pratap Singh was the original leader of the *Zehreeli Gas Kand Sangharsh Morcha* and now manages an umbrella group of NGOs providing a range of services from economic rehabilitation for young people from the gas affected communities, to HIV/AIDS and public health awareness. An educated outsider from a politically influential family, who speaks English as well as Hindi, he was one of the first of the outsiders on the scene when the need was most acute. For Alok, the priority is practical relief for the victims; by the government if possible, with social protest achieving concessions where necessary, and by NGOs or direct action if not. Pragmatism over-rides idealism. It is important to accept compromise in order to deliver material benefit to the victims. In the early days after the disaster it was important for *Zehreeli* to campaign to express unmet demands, to achieve rights and basic services. Now there is little to campaign for. What is needed now is practical economic rehabilitation schemes to get jobs for the poor. The role of intellectuals is to organise the poor because that gives best delivery to the poor, until such time as the poor have their own level of organisation.

Abdul Jabbar Khan is a local man from a poor background with a moderate level of education, who ran a business before the gas leak and became secretary of a community *mohalla* committee in the early days of the disaster. He was invited to advise the *Bhopal Gas Peedit Mahila Udyog Sangathan*, Women Worker's Union and later became its leader. For Jabbar it is only the poor that can lead the poor. It is important to have a mass following which you get by delivering what people want, which is primarily compensation and jobs. People's loyalty to the movement is built by independence, in particular financial independence. Take no funds from outside, raise it all in small subscriptions from the masses or from government grants so long as it's locally controlled.

Balkrishna Namdeo was from a poor background and moved to Bhopal as a child for the purposes of education. Communist by inclination, cultivated, moral and somewhat ascetic, he has never married and has devoted his life since before the gas leak to working for the poorest and most vulnerable, those who are entirely dependent on the state for their survival. He expects and demands high standards from the state which he patiently pressurises to deliver their responsibility. The widows, pensioners and severely disabled with whom he works are devoted to him and look up to him for help and direction. Gentle and kind, he sees his role as leading and providing for those in his care.

Satinath Sarangi ('Sathyu') is a principled idealist. From an educated, English speaking, left wing background, he was one of the earliest outsiders to arrive in Bhopal after the disaster. He operates on a global stage and has done much to internationalise the campaign. His ideological position prevents too much compromise which some find difficult to work with, and consequently he has often fought a lonely battle. For Sathyu, the battle is never over until every last multinational has been brought down by the power of the poor – this is the goal and much must be sacrificed for this. Tactical rather than pragmatic, he is devoted to the poor but willing to use whatever means are necessary to support their interests, including the money and expertise of the rich west. The only progress possible is that which can be controlled by the people: industrialism is out, small is beautiful; voluntarism is all. Always demand more of people to achieve collective self-management. Neo-Gandhian / Anarchist.

Sadhna Karnik arrived in Bhopal after the disaster as a young, educated woman. For Sadhna compassion and discipline drives service: service to the survivors and service to the communist party which she was drawn to through her activism and which has a coherent programme within which her work has relevance. The ideological battleground leaves only a few big (male) warriors clashing and those with genuine compassion are left battered and hurt.

In order to retain their effectiveness, campaign leaders must attract loyalty from their followers and demonstrate influence with those with the power to make changes. Beside the ordinary human temptations which go with successful campaign leadership: power, influence, fame, adulation, honour, access to resources not to mention the offers of bribes from those in power, there is always a tension between short and long term goals, between process and outcome, between results and principles. The broader the coalition of supporters, the more difficult it is to maintain loyalty. With a situation as complex as Bhopal, with its legal, medical, environmental, political and global precedents, this becomes even more difficult as strategic and tactical decisions are made in entirely new territory.

To outsiders, listening to the positions of the protagonists in the movement, the differences and divisions between the groups appear to reflect the major dilemmas and tensions of any community action or social movement. What are principled positions inside the movement appear as differences of tactics to the outside. Many dialectical tensions are evident in the Bhopal movement which are present in community organising and movement politics anywhere. The Bhopal survivors' movement is an encyclopaedic microcosm of the politics of protest and community organising, albeit within a particularly Indian context.

From whom is it appropriate to receive funds? There is a tension between the purity and resultant austerity of taking no funds from outside and doing what can be done with the resources of the membership and local supporters, versus the opportunities of what can be done with the resources of wealthy supporters, with the associated potential for complacency, dependency or even corruption which this brings. At one extreme, ICJB receives funding from foundations and individuals across the world, works with paid activists and finances big protest actions. At the other end of the spectrum, *Pension Morcha* operates on a shoestring from Namdeo's house, with costs covered only by local donations from the poorest. And what are the implications of this? In order to continue receiving international funds, groups need to demonstrate to an international audience the value of its work and there becomes a tension between international awareness raising of the issues, and international fundraising for an organisation. Such tensions may be creative, used to the benefit of the movement, but the risk remains that they lose their footing and shore up problems.

Decisions about receiving money from the state in its various manifestations forces the question of relationship to the state: the democratically elected yet to varying degrees corrupt institutions of government; the ponderously slow legal process with differential access but in principle, equal treatment of rich and poor; the daily contact with bureaucrats and state functionaries with their own interests and more than fair share of crooks. ICJB's foreign funding allows it to remain at a distance from the state whereas BGPMUS, through its training wing *Swabhimaan Kendra* has found ways of making positive use of state money consistent with the union's principles. One of the *Stationery Sangh's* earliest successful campaigns was for the nationalisation of the stationery workshops and the improved terms and conditions which go with direct state employment. For *Pension Morcha*, the state is the primary focus for campaigns as provider of last resort to the economically inactive. Any state, but especially a liberal democratic and multi-level post-colonial state such as India, contains within it contradictions between roles: as a medium for redistribution and public service provision, open to influence through the democratic process whilst at the same time a self serving bureaucracy and protector of ruling interests with a near monopoly of legitimised coercive violence. The groups have orientated themselves in varying positions with respect to this complexity and this is reflected to some degree in the interviews with activists.

Relations with the state also throw into relief issues about organisation. If not within the state, what is the appropriate form of organisation: NGO; Trust; trade union; membership organisation; trading company; cooperative; collective; coalition; informal association? How is the tension resolved between popular, open democracy and effective decision making by those with the skills, experience and expertise? This is even more pronounced in a context where literacy is confined to a minority of disproportionately male activists in a largely female movement; where trust of individuals often has greater currency than documentation and accounting conventions; where technical knowledge of chemical, medical, legal, commercial and political issues are essential yet used by the powerful to exclude survivors. How knowledge – technical, experiential and indigenous - is obtained, translated and used for the benefit of the campaign has been an important subject of study in this research.

The practice of political action inevitably involves compromise, engaging with a system in order to change it. Exactly where these compromises are legitimate and what are the bottom lines remains a source of strategic tension. In principle, everyone agrees that the continued presence of a contaminated factory site 25 years after the disaster is a disgrace, and that Dow, as successor company to Union Carbide, is responsible. But in practice there is a tactical decision whether it is better to hold on a bit longer for the leverage it gives to demand Dow's accountability, or for the state to conduct a quick (and dirty?) clean up and chase Dow later. In principle, all agree that Bhopal is not a unique case and that economic and industrial policy in India needs to change, but the extent to which that means adhering to a political programme, and the collective discipline that involves, as against extracting piecemeal reforms from whomever can give them even when this leads to political promiscuity and unsavoury liaisons, is a matter of judgement.

Social movement theorists refer to interpretive 'frames' which enable campaigns to make sense of their own and others' activities, to help interpret events, discern allies from enemies and ultimately to engage with the culture of society which it is hoping to influence. These frames change with time and the work of developing tactics, negotiating concessions, building alliances and outflanking enemies. They are shaped in the interaction between the campaign's ideologues, supporters and those who they influence. Arguably, the groups in the Bhopal movement reflect an environmental justice frame, an employment rights frame, a service provision frame and a communist programme frame. These frames overlap and many campaign objectives, such as the demand for compensation, are common to all. However there are times when frames are stretched too far with the result that groups split, individuals fall out and even feuds start.

Aspects of these dilemmas are well reflected by Tarunima and Dharmesh.

Tarunima: On my 2nd day I saw candles being lit in the shape of 'NO MORE BHOPALS' in English. But it struck me that there was no Hindi reference to the same sentiment. Yes, those active in and working for the *sangathan* knew what it stood for but what about those who had gathered there? Wasn't the movement of the people and not a selected few? When I questioned people I got two different answers, both of which sounded valid enough but are a bone of contention for the other *sangathans* in Bhopal. First is the large number of international media present during the anniversary and their presence being stronger and better than national media. Secondly is that it portrays a truth that one cannot ignore but the possibility of lighting candles in the shape of another slogan would have been difficult. But the other *sangathans* question why it is necessary to use English. Why not the language that is locally spoken?

The same could also be said for their dissatisfaction and irritation with a musical band performance on the 3rd December. It was performed by a local college band and although they sung in Hindi, the genre was rock. The lyrics were surely heart felt but the number who knew what they were singing was few. Here, too the other *sangathans* find it difficult to associate. They feel that all this is for "outsiders".

For them, because the publicity is being done in English and the lyrics of a song being lost in its unknown music, it does look like an act put together for the foreigners. On the other side the ones who have their "outsiders" on their side to support and to keep them going feel that the local *sangathan* is asking the people for their hard earned money to run a campaign which does not show any evidence of money spent or collected. This being one of the basic conflicts, finding a neutral position and taking a neutral stand was not an easy task and then to prove this was even harder.

Dharmesh: Due to the dynamic nature of the movement and several ideological beliefs, groups within the movement have a history of alliances and break ups. From a researcher's perspective this was more of a resource than a problem as it helped in understanding the diversity of approaches. These ideas had either evolved or in some cases were inherent among members. For instance, funds raised from foreign donors was a volatile issue, with arguments for and against it fairly established. Legal suits filed during previous alliances sometimes brought groups together to discuss strategies. Issues like the clean up of the abandoned factory site brought out differences over the liability of the issue. For the International Campaign for Justice in Bhopal the acceptance of liability for clean up by Dow Chemicals is based on a larger goal of setting a global precedence while *Zehreeli Gas Kand Sangharsh Morcha* believes that the immediate clean up and not assigning liability is priority for practical purposes.

Themes

One of the principal objectives of research was to identify key themes that emerge from the Freirean process of interrogation of codes. Good themes in this sense are consistent with the thematic universe of the survivor-activists in the movement yet are also generative of new insights which are really useful to furthering their collective interests. Generative themes stimulate learning as part of the cycle of action and reflection which is movement praxis. These are inevitably contingent, subject to testing and accountability to the movement in a process which is constantly ongoing. However, a selection of themes which have been presented to movement activists from all existing campaign groups in a public symposium are given below:

Gender

This has been a key concept that defines an industrial disaster of such magnitude. To begin with women were the most vulnerable victims of the gas leak, both in terms of the breakup of the family unit, problems of reproductive health and the social ostracisation that followed. They were twice victimised – as marginal members of a patriarchal system, and victims who faced the brunt of the aftermath of a disaster. They were virtually erased from the official discourse yet became the embodiment of an alternative narrative, from the carnal experience of pollution to the tactics of bodily confrontation[10]. Today, they are the most recognisable faces of the struggle for justice in Bhopal. What are the motivating factors and how have women grown from *burkha* clad, domesticated entities to group leaders? Gender becomes an important category for tracing such a growth both in terms of the cognitive aspect of a social movement and its practice. Gender discrimination in government policies is rampant and this has influenced social attitudes, so that women have had to take up cudgels at the public and personal level. Our interviews show how sharply their learning graph has grown over the last 25 years, and how the different survivor organisations have learnt to make their demands gender sensitive. In many ways Bhopal has broken gender stereotypes and offers a case study of the how individual growth and collective strength can rupture hegemonic ways of determining knowledge by the privileged, ruling class. In learning to know and control the woman's body it becomes the site of resistance. At the same time, what has happened to the largely absent men? A handful has influential positions in the campaign groups and their responses to the questions of leadership and decision making interact with how their own masculinity is seen, by them and others. A few more participate in actions from the rank and file, or support from the sidelines. Also hidden amongst our interviews are those invisible men who quietly, in contradiction to prevailing attitudes, give encouragement to wives, express pride in mothers, make sacrifices for the education of daughters, and at the same time are crafting new ways of being men.

Gendering the movement also provokes questions of family dynamics. The women survivors tell of a diversity of experiences of resisting and accommodating to domestic patriarchy. Unusually high levels of widows, divorcees, separated, remarried and unmarried women, single mothers and women-headed households exist alongside enforced purdah, domestic abuse and violence. However, amongst women, the experience of empowerment cohabits with the desire to reinstate patriarchal endogamy for the next generation. This is not a simple, linear story of feminist enlightenment but a complex social reality of renegotiating the gender regime as women struggle for dignity and justice.

Religion

North Bhopal, where the Union Carbide factory is situated, has a high Muslim population and survivors are Hindu and Muslim in approximately equal numbers with only a handful of members of other religious communities. Religious leaders are not active in the movement and have tended to side with official presentations of hand-wringing public mourning. Religious organisations, at least since the immediate post-disaster provision of welfare, have been absent, with the only exception being low level support from international Christian ecumenical organisations (eg World Council of Churches) and a few local nuns. North Bhopal is however a visibly (and audibly) religious community, the regular call of the muezzin interspersed with the sound of *pujas* and ceremonial drumming.

Within the movement, many religious taboos and practices are infringed, such as *burkha* wearing and intercommunal and intercaste food sharing. People trace the origins of these infringements to the immediate aftermath of the disaster, where in the urgency of escape, traditional notions of propriety were abandoned, and scarcity of food and water demanded that questions of origins became irrelevant. However the changes to religious practice were not simply abandoned in panic. Our interviews reveal stories of careful thinking and interrogating of sacred traditions and testing them against the reality of struggle. This is not a simple process of secularisation of activists, who often profess continued if not increased devotion to their faith as a personal resource or source of motivation and hope in the struggle. On the contrary, it might be suggested that what is happening is a grassroots renegotiation of religion from the bottom up, in an interfaith struggle for humanity in a new post-disaster social reality.

The state

There are conflicting narratives of the state. Many survivors compare their campaigns for justice to the national liberation struggle for independence from British colonialism. Expectations expressed by survivors often correlate with the Nehruvian ideal of big, paternalistic state - adequate health care, provision of employment, protection of the weak – of which it is found wanting. Moreover, everyday experience of the state involves petty bureaucracy, corruption and brutality which has denied to many access to what little compensation has been available. The state and government is blamed for the gas leak more harshly than Union Carbide or Dow Chemicals. Since the 1990s, the Indian state has adopted aggressive neoliberal policies, a conciliatory approach to multinationals such as Dow and using security repression to quash protest.

Knowledge

The majority of survivors have very limited education and literacy. Many aspects of the campaign draw on technical knowledge, detached from the experience of those affected. Activists are dependent on sympathetic intellectuals but demand accountability. The movement also generates its own, lay knowledge through experience and activism. Knowledge is generated through the interaction between campaigning practice and sympathetic expertise, and is translated through diverse channels of communication in different campaign groups, between leaders and affected communities and with metropolitan and international elites with power to affect change.

The remainder of this book constitutes the stories told by the survivors and those who have been active in the campaign over the 25 years. We start with an essay written by Alok Pratap Singh, the leader of the original *Zehreeli Gas Kand Sangharsh Morcha*, which formed from the activists and supporters who arrived in Bhopal immediately following the disaster, stimulated the formation of the *Mohalla* neighbourhood committees and which became the first umbrella organisation for the survivors' groups and supporters for the first few years of the struggle. Singh remains in Bhopal, managing an NGO which provides economic rehabilitation to survivors and other services in the city. Here he documents the early days of the movement and the crucial legal interventions which the *Morcha* has made since.

Singh's essay is followed by a series of pieces which describe the formation, early development and subsequent work of the biggest of the survivors' trades union organisations the *Bhopal Gas Peedit Mahila Udyog Sangathan*. This starts with an excerpt from an interview with Rabiya Bee, a worker at one of the work sheds established for economic rehabilitation of women survivors, and the founder and original convenor of the trade union which formed to protect the jobs and improve the working conditions of the women. This is followed by an interview excerpt with Mohini Devi who joined later and is currently in a leadership position. The leader of BGPMUS since the early days is Abdul Jabbar Khan, and an essay by him forms an account of the union's philosophy and how it remains distinctive in the movement.

This section concludes with two more interview excerpts, one from Hamida Bee, another influential figure in the movement currently, who recounts some of the key achievements of the sangathan, and Rehana Begum, a former activist who remains critical of some of the directions the organisation has taken.

Following this section are two interview excerpts and an essay from activists associated with the International Campaign for Justice in Bhopal. Syed M. Irfan was a trade union organiser who became active in the early *Mohalla* neighbourhood committees, later joined with the BGPMUS and then formed *Bhopal Gas Peedit Mahila Purush Sangharsh Morcha* which remains one of the key members of ICJB working at grassroots level. Another ICJB member organisation is the *Bhopal Gas Peedit Mahila Stationery Karmchari Sangh*, a trade union formed at one of the smaller economic rehabilitation work sheds manufacturing stationery products. Rasheeda Bee, one of the founder members and leaders since its inception, recounts the story of the union's early years from 1986 onwards, the struggles for job protection, nationalisation of the worksheds and equal treatment with other government employees and how it has subsequently taken on wider and more environmental issues. This section is completed with an essay by Satinath (Sathyu) Sarangi, an original member of the *Zehreeli Gas Kand Sangharsh Morcha* and originator of Bhopal Group for Information and Action, a solidarity group which has played a central role in the ICJB. Sarangi is also Managing Trustee of Sambhavna Trust, an internationally recognised clinic providing health services to survivors.

The third of the main campaign groups, the *Gas Peedit Nirashrit Pension Bhogi Sangharsh Morcha* is described in two pieces, one an essay by the group's leader, Balkrishna Namdeo, who describes his own background and philosophy as well as the struggle for pension entitlement and conditions, and an interview with Badar Alam, a young activist in the group.

The section which follows fills in some themes which are not picked up elsewhere, and those which are articulated by rank and file activists.

Hajra Bee was initially involved in a rehabilitation work shed union associated with the Communist Party and no longer in existence, and more recently became active in ICJB. She argues that many Muslim women like her have stopped wearing the *burkha* although this does not entail a rejection of the principles behind *burkha* wearing. An essay by Sadhna Karnik demonstrates that the Communist Party continues to be an important source of solidarity for the survivors. Om Wati Bai has experienced workplace hazards since before the gas leak and her grandchildren continue to suffer. She describes herself as a follower in the movement, not a leader, and is equally willing to participate in the actions of any of the various campaign groups. Razia Bee and Ruksana Bee, interviewed together, formed part of a breakaway union when the *Bhopal Gas Peedit Mahila Stationery Sangh* split (temporarily) after disagreements with leadership. This interview demonstrates the tension within a movement which continues to hold together the day to day needs of workplace conditions with the wider principles of environmental justice. Another self-proclaimed follower albeit with a key role in ICJB is Nawab Khan whose interview gives a philosophical angle on the future of the movement.

The future of the movement lies in the next generation of activists. Children have been greatly affected by the gas leak, whether directly gas affected, orphaned or through second generation congenital problems. Shahid Nur, orphaned by the gas disaster, initiated in his young adulthood a campaign group of fellow orphans who for a while represented this important group of survivors. Sarita Malviya, interviewed in 2007 aged 14, has become a leading advocate of youth activism, especially amongst those who continue to be affected by contaminated water. Educated, articulate and angry, the young people of Bhopal are taking receipt of the baton of responsibility for the demand for justice. The recently formed Children Against Dow Carbide, interviewed just after their first protest rally in 2009, make it clear that 25 years on, the struggle for justice in Bhopal continues.

End notes

1 In nineteenth-century Britain, a radical working class education movement rejected the philanthropic 'Society for the Promotion of Useful Knowledge' on the grounds that the 'usefulness' of the knowledge was judged by the ruling class and designed to stifle political dissent. In parody of the Society, the radicals advocated 'really useful knowledge' as judged by the movement, and included the study of Thomas Paine's *The Rights of Man* and Karl Marx's *Communist Manifesto*, works which were excluded from the Society's syllabus.

2 Full, anonymised transcriptions of interviews in Hindi and translations in English will be available in print in various locations and on-line from Queen Margaret University archive on http://edata.qmu.ac.uk.

3 Norman Denzin and Yvonna Lincoln have compiled and edited what is known to many as the 'bible' of qualitative sociological research: *Strategies of Qualitative Inquiry* (Sage, 2003). In addition to the main metaphor of quilt making, they also describe such research as like jazz musicians who improvise melodies, harmonies, rhythms and instruments in interaction with one other, interpreting each others' moods and intentions and collectively producing a piece of music which surpasses the sum of the parts. As a sociologist who has never made a quilt but enjoys jazz, Eurig prefers their jazz metaphor.

4 Michel Foucault, in *The Archaeology of Knowledge* (London and New York, Routledge Classics, 1969) analyses history from the perspective of the shifting layers of discourses which exist amongst the public and the powerful of any period.

5 Many of the factual details have been gleaned from Bridget Hanna, Ward Morehouse, and Satinath Sarangi, ed. *The Bhopal Reader: Remembering Twenty Years of the World's Worst Industrial Disaster* (Goa: Other India Press; New York: The Apex Press, 2004).

6 See Upendra Baxi, and Amita Dhanda, *Valiant Victims and Lethal Litigations: The Bhopal Case* (The Indian Law Institute. Bombay: N. M. Tripathi Pvt. Ltd., 1990).

7 Material drawn from *Clouds of Injustice: Bhopal Disaster 20 years on* (Oxford: Amnesty International, 2004).

8 Greenpeace *The Bhopal Legacy: Toxic Contamination at the Former Union Carbide Factory Site, Bhopal, India, 15 Years After the Bhopal Accident* (November 1999).

9 *Health Effects of the Toxic Gas Leak from the Union Carbide Methyl Isocyanate Plant in Bhopal: Technical Report on Population Based Long-term Epidemiological Studies* 1985-1994 (New Delhi: Indian Council of Medical Research).

10 Suroopa Mukherjee, *Women Survivors of the Bhopal Disaster*, forthcoming from Palgrave Macmillan as an imprint of the *Palgrave Studies in Oral History* in 2009.

Alok Pratap Singh

Zehreeli Gas Kand Sangharsh Morcha
Poisonous Gas Episode Struggle Front

Rajeev Smriti Gas Peedit Punarwas Kendra
Gas Victims Rehabilitation and Training Programme

Alok Pratap Singh was founder and convenor of the original campaign group to form in the aftermath of the disaster *Zehreeli Gas Kand Sangharsh Morcha*. He is founder president of *Rajeev Smriti Peedit Punarwas Kendra*, an umbrella group of NGOs providing a range of services including economic rehabilitation for young people from the gas affected communities. He is a member of the Coordination Committee on Bhopal (Ministry of Chemicals, Government of India) and Sub Committee on Economic Rehabilitation (Department of Bhopal Gas Relief and Rehabilitation, State Government of Madhya Pradesh)

The *Morcha* and the engineering of a mass movement

In the immediate aftermath of the gas leak, a rescue and relief operation was initiated by local social activists, students, youth, trades unionists and others who were already working for the development of different sections of the poor of Bhopal. Many activists from different corners of India came to Bhopal to support the relief effort and join the struggle for adequate services, amongst whom were prominent educationist Dr. Anil Sadgopal and others. With the formation of the *Zehreeli Gas Kand Sangharsh Morcha* ('*Morcha*' (front)) on 7th December 1984 we decided on the following lines of action: 1. to save survivors; 2. to struggle for relief rehabilitation and medical care; 3. formation of people's committees at grassroots level; 4. mass awareness and education to the victims for their right to life; 5. survey and data collection; 6. building up national and international support networks; and 7. legal interventions. The principal slogan of the *Morcha* was evolved:

"Struggle for peoples' rights, people's science, people's unity."

After the formation of *Morcha* many people have joined our struggle including Satinath (Sathyu) Sarangi, Sadhna Karnik and Abdul Jabbar who are still working among victims but with different organisations.

Most of us already had experience working with various forms of mass movements before the gas disaster and had used the local committee structure, where key people were identified and given responsibilities at grassroots level. We were aware that these people are the real backbone of any mass movement, and sometimes better than us or any government official, to understand the complexity of the situation. We knew that this was going to be a long struggle and it was evident that it had to be done by the organised, conscious and disciplined local people. We had decided during the inception of the *Morcha* that this would be a people's movement: it would be fought by the organised people and the mechanism would be designed by the people with our assistance. At first there was no discrimination based on ideologies or political background, even though we were from different organisations and movements.

The Government initiated 'Operation Faith' (neutralisation / utilisation of remaining gases) and set up relief camps on December 16th 1984. They then decided to close the camps after a few days without caring that the priority was to provide food, water, and basic rations to the victims until socio-economic conditions were back to normal. There was a spontaneous outburst from the people in the relief camps against the Government's decision. The first demonstration of victims was led by us at *Raj bhavan* (Governor's House) on 18th December and the state government was compelled to establish centres for free distribution of basic rations, food, milk, tea, cloth etc. at neighbourhood level. Recognising the need of the situation, we immediately organised people's committees at neighbourhood level to make the distribution system smoother by ensuring effective people's participation It was the beginning of the people's committees as an instrument of struggle. With the help of these committees we started surveys of people in the affected area and identified people in these *bastis* (settlements / neighbourhoods) who had the skills to take charge of the issues faced by their community. Further, these people helped to form the *mohalla* or local committees. We had managed to form such committees in 23 of the *bastis* by the end of December.

Historical people's march: A turning point

All the important decisions were made in consultation with all the committee members during meetings at the *Aloo Godown* (disused potato cold storage facility) near J.P Nagar. In December 1984 at the first meeting of all the *mohalla* representatives, it was decided that we should restart the survey that the government had conducted which the people were not satisfied with. On the 23rd December we released a survey proforma and started the surveys. This was around the same time that the government surveys were being conducted by the Tata Institute of Social Sciences (TISS). Our own survey was launched to gather information that was left out by the TISS Survey and also to answer questions regarding availability of all kind of relief announced by the Government, ration card, medical relief, deaths of family members whether surveyed by TISS or not, cattle loss etc. The people's survey was done by local educated people who were members of the committees. This ensured that we had good results and more accurate information. The results of these surveys were supposed to be used to motivate the Government to design their policies with the right perspective.

In the process of surveying, problems of the victims had been identified and a people's charter of demands emerged. We recognised the deep pain, agony and anger which was agitating the minds of the mass of the victims, and they were becoming restless. The meeting of 32 *mohalla* committees decided to hold a massive demonstration of victims before chief Minister of the State on January 3rd, although this was not supported by the 'outsider activists'. In this process almost all organisations involved in relief work taken into confidence including *Rahat Aur Punarvaas Samiti* (Relief and Rehabilitation committee) which was formed by veteran Communist leader B.K. Gupta, a film maker Tapan Bose and film actress Suhasini Mule of Mumbai.

Suddenly, *Rahat Aur Punarvaas Samiti* announced it would hold a *dharna* (vigil) on bus stand square on 1st January. Obviously, it was a move to disrupt our decision. However, we decided to participate in that *dharna* with full strength rather than watch it from the fence by overruling the opinion of 'outsider activists' again. *Dharna* was successful but unplanned. It started with the commitment that it would be called off only once all the demands were met – which was a great mistake. We all knew that this was an unrealistic, premature commitment. Eventually the leaders were forced to call off *dharna* unwillingly and there was a big public reaction against leaders and people were feeling let down and angry. At this point it was necessary for the people to get motivated in a positive direction. We stood firmly with our prior decision. On 2nd January we held a big meeting at the *Chhola Naka*, of all the community groups that had formed by then and to which all kind of organisations and political parties were invited. We ensured that the local community groups were not taken over. No matter how small they were it was important for them to keep their identity.

The demonstration of 3rd January justified our trust in the masses and their creativity. Beyond our expectation a big human sea turned out on roads in front of UCIL plant. It was the first organised demonstration of gas victims in the history of the movement. About 12,000 to 15,000 people joined a 2 km long protest march which ended at Chief Minister Arjun Singh's residence, and it was followed by a 7 day *dharna* attended by more than 5000 victims per day. There were a lot of negotiations and side by side there were attempts to break the movement too. Most demands had been considered, deadlock remained on some important demands and we had exhausted this form of tactic. *Dharna* was called off on 10 January to enter immediately into a more militant form of resistance: "*rail roko*" (stopping the trains by blocking the tracks) on 12th January. A lot of leaders were careless and got arrested by the police just before the demonstration but the protest still managed to go ahead.

Emergence of a national level network

In order to get more national and international attention and to create a national level support system we held a meeting of NGOs, medical organisations, trades unions, student, women and civil liberty & democratic rights organisations in Bhopal from all over the country. A National Campaign Committee for gas victims consisting of 144 organisations was formed. I was on the steering committee, and so was Anil Sadgopal and Vibhuti Jha. This brought in a national perspective that the campaign needed. Members of National Campaign Committee decided to carry out various programmes in different parts of the country in support of the Bhopal gas victims. Medical organisations like Medico Friends Circle (MFC) Mumbai and Drug Action Forum(DAF) Kolkata also joined. Our local theatre group held more than 150 shows of street corner play *"Khamoshi tod do"* (Break the silence) in different parts of the country exposing linkages between UCC/UCIL with Government of India and State Government of M.P.

Then in April 1985 we held a big demonstration, joined by 5-6,000 victims, at the residence of newly elected Chief Minister Motilal Vora, for demands and suggestions based on the people's survey, about medical care, socio-economic rehabilitation and relief to victims. Pressure of this mass upsurge forced Government to act on demands of our memorandum and to start working seriously on all these major issues. In every municipal ward, sector offices were opened to monitor relief work, survey, and speedy disposal of *ex gratia* payment for deaths. Announcements were made that four new hospitals would open including 200 bed hospitals, pulmonary centre and many dispensaries. However, whilst most of our demands had been considered, the reality on the ground was that government officials were creating all the hindrances which were inbuilt in the system.

People's health clinic and programme

Morcha had conducted powerful health actions with Medico Friends Circle (MFC) from Mumbai, and Drug Action Forum (DAF) from Kolkata including health surveys and awareness campaigns. We approached the Supreme Court of India to direct the state government to use disposable syringes to save victims from hepatitis epidemic, which it was forced to do, and also to provide injections of sodium thiosulphate (Nats). Nats was shown to be effective in detoxification of victims of Cyanide-like poisoning but the state government refused to provide them. Manufacturing of Nats had started at IDPL (Indian Drug Pharmaceutical Ltd.) Hyderabad. However, the Health Director in the state government, who was known to be close to UCIL, had discouraged the administration of Nats at hospitals. That forced us to establish an alternative People's Health Clinic (PHC) to administer Nats injections ourselves with medically qualified volunteers. On 1st June 1985 gas victims forcefully entered UCIL premises and started a detoxification centre to administer sodium thiosulphate injections and carry out detailed monitoring of its effect. This was the *Jan Swasthya Kendra* (People's Health Clinic). It was also symbolic of people's control of the UCIL plant. Dr. Mira Sadgopal, Dr. Nishith Vora (MFC), Satinath Sarangi and Sadhna Karnik (*Morcha*) took charge of PHC. The clinic was however forcibly closed by the police on 25th June 1985 and all the data collected by us, which were essential for the health of the people, was confiscated. This was part of a new wave of brutal repression by the state.

Brutal suppression of victims

The first session of the newly elected State Legislature Assembly was scheduled for June 1985. We took it as an opportunity to raise the issues in the house with the help of Assembly members, most of them known to me personally because of my political background and my family's history in the freedom fight. On 25th June a massive flash protest was planned in response to government apathy but the more inexperienced leaders were careless and got themselves lifted by police late the previous night. I was the only one who escaped arrest. Using the second and third line of leadership the message was successfully communicated to local committees and especially to women volunteers. When I reached J P Nagar from the rear entrance the people were already waiting for me to arrive. There was heavy police security in the entire area. I started announcing the rally and began to bring people out of their homes, and within a few minutes 3,000 women had gathered. As they began moving ahead more people joined and in the end the rally became nearly 9,000 people. The idea was to keep me in the centre of the crowd to avoid any police arrest.

All the other leaders were detained in the control room so the rally headed there first to get them released and then headed to the *Vallabh Bhavan* (Secretariat). We went there to meet the Chief Minister but he had already fled to Delhi so no policy decisions could be made. The Chief Secretary called in a delegation of 35 people for discussions. There was a lot of dissent amongst the delegation with Satinath and Sadhna demanding much larger things which were not in the remit of the Chief Secretary nor were they in the immediate radar of the victims who were part of the delegation or those who were waiting outside. Finally the Chief Secretary agreed to direct the Collector to sort out within a week all of the most basic demands of the people like surveys and medical care.

When we returned downstairs to talk to the people all we saw was a sea of footwear. There had been a heavy *lathi* charge (attack with large wooden sticks) on the people by police. The movement had been tricked. Victims were beaten badly wherever police found them. We later learned that some 'outside activists' from Delhi who often come for 'tragedy tourism' or adventure experiences had, instead of waiting for the delegation, provoked the victims and thrown stones on police and this had led to a natural extreme reaction from the police. It was a move for which people were unprepared. It was the stage of a movement which one would expect to take only propaganda form. Victims were not educated and prepared for the movement to go into a phase of a higher form of violent resistance which had been the consequence of this mindless provocation.

This was very bad for the victory that we thought we had achieved. If it had not been for this negative incident we had almost won our other demands, because we had the masses with us to pressurise the government and we had nearly 50 MLAs supporting our demands inside the Legislative Assembly for the next proposed step – "Encircling assembly by human chain". After this incident we also lost a lot of supporters. One of my friends, Vijay Dubey, MLA of the ruling party even linked us to the CIA. We all were arrested, including 'visitor activists'. We decided not to bail out from jail, but this move failed when all 'visitor activists' decided to bail out, full of tears. About 30 cases were filed against us which were ultimately withdrawn by the Government in 1996-97.

Turning the movement around: New challenges

Suppression of victim's movement and our arrest was severely condemned nationally as well as internationally. It was a set back at local level but big support emerged at national and international level. After 7 days of judicial custody, we were immediately involved in a damage control exercise. An open letter to the Chairman of the Legislative Assembly and its Members was released. Sathyu and other activists threw pamphlets in the house of the Legislative Assembly. An urgent meeting was called of the National Campaign Committee and *dharna* was planned in Delhi on 24th July. On August 30th "Throw carbide out week" was planned 27th November to 3rd December. A petition with 4 *lakh* (400,000) signatures was submitted to Union Government in early 1986 during a victims' rally at Delhi.

In 1986, activists and groups from outside slowly started withdrawing from active participation from the movement. It was not physically possible for them. Sathyu formed Bhopal Group for Information Action (BGIA) because he was not allowed to separate PHC from the control of the people's movement. Now he manages some good foreign funding for health projects in Bhopal. Anil Sadgopal returned to Hoshangabad. Vibhuti Jha returned to his previous legal profession at Bhopal District Court from where he took up our legal struggles. Most of our supporters returned to their work. Victims were involved in accessing all the relief which had been achieved through struggle. Fear of police suppression still prevailed among victims. Sadhna Karnik and I remained in *Morcha* with the victims and their shattered committees. We managed to run PHC again in a rented place and detoxified more than 3,000 people. We were short of funds, which had previously been arranged by Anil, Meera Sadgopal, and Sathyu and were now diverted to other ends. This was the time when we were in search of new, educated and committed people, like Abdul Jabbar, who joined the *Morcha* and later organised 2,000 women as an independent trade union at the sewing centres which had been opened by the state government under a programme of economic rehabilitation,. This union was headed by him and helped by Anil Sadgopal and Sathyu as a rival organisation to the *Morcha*. This was the period of all-out crises for us.

The legal battle: Interim relief, a great victory of the victims

When the legal case was sent back by the US court in May-June 1986 I saw a big light of hope. After discussion with Vibhuti Jha we agreed that victims should be awarded interim relief because the compensation case might take a long time. We did not expect the Government to file this application so we decided to go ahead with it. We filed two applications in the court on behalf of both *Zahreeli Gas Kand Sangharsh Morcha* and *Jan Swasthya Kendra* (People's Health Clinic): one for the interim relief and another one to make us the interveners. Justice Deo accepted the applications. The argument for interim relief was that since there was no dispute over the basic facts of the disaster (there is a gas leak and there are victims) so there was no impediment to awarding interim relief before the case had been completed. The concept was introduced by us for the first time in June-July 1986. At the time hardly 400 people were active in the *Morcha* but we were surviving. Vibhuti was a fresh lawyer who had only 4 years legal experience whereas UCC and the Government had big lawyers of the Supreme Court of India and High Court of Madhya Pradesh in Jabalpur to represent them. But we trusted Vibhuti because he was one of the founders of *Morcha*, a close ideological friend and committed to fight the case without fee. Our lawyer was not heard so we had to find a different tactic and held a demonstration at each hearing. This attracted a lot of media which led to an increase in participation and created new hope among victims. At the last few hearings we were supported by various groups, political parties and organisations.

On 17th December 1987 Justice Deo of the District Sessions Court ordered an interim relief of Rs.350 *crores* (Rs 3.5 billion. Approximately £35 million at that time). In the entire history of the Bhopal legal case this has been the only judicial order that has been in favour of the victims and against UCC. Union Carbide appealed against this order in the High Court. On 4th April 1988 by the order of High Court the amount was brought down to Rs.250 *crores*. Then the case went to the Supreme Court in July–August 1988. On 14th February 1989 the Supreme Court passed the order for the 'out of court' settlement.

Immediately after the settlement was announced, numerous organisations agitated against the Supreme Court demanding the settlement be quashed. We were against quashing the settlement: we agreed that the amount was very little but it was not a wise thing to quash the order which would mean that the money would go back to UCC whilst the victims needed that money desperately. Instead we filed another petition for more compensation and for a mechanism by which the money can be efficiently distributed. We also fought that direction of the Supreme Court which aimed to free UCC/UCIL from its criminal liability.

Soon after, the central government led by Rajeev Gandhi was replaced by a coalition government led by V.P.Singh. This government announced Rs.7200 interim relief at Rs.200 per month for three years among the victims of the most seriously affected 36 out of 56 municipal wards, and refused to sanction the long term Action Plan of the previous administration.

New fronts: Economic rehabilitation and removal of toxic waste

By 1990-91, all the sewing centres established by the state government had been closed because of growing trades unionism. In 1995, a newly elected state government decided to run all these centres with the help of NGOs. As a by-product of *Morcha* we formed *Rajeev Gandhi Memorial Gas Victims Rehabilitation Centre,* as an apex body with the network of 10 NGOs and in memory of late Prime Minister Shri Rajeev Gandhi whose Action Plan was never carried out. The state government provided us infrastructural support and we managed to run these centres successfully with the help of many schemes of central and state government. These centres are still working and have benefited more than 20,000 families in these years.

Some of the groups had filed applications in 1999-2000 to clean up hazardous chemicals from UCIL factory in the American courts, in order to pin the responsibility on UCIL/Dow chemicals. Nothing had come out of it in four years. We filed our petition to clean up the site in Jabalpur High Court in June 2004 and the Court gave a verdict in 9 months. The Indian Courts have been quicker. There might be problems with our laws but for Bhopal, all the courts of India have provided quicker justice than any American court. The Court said that all the issues of liability against Dow are valid but our priority at the moment is to save the people who are being poisoned. The Government can pay for the clean up and later claim it from Dow Chemicals. It was the Government's responsibility as the *Parens Patrie* according to the Bhopal Act. According to the order of the High Court the chemical waste should be removed and disposed off appropriately at Ankleshwar in Gujarat state or Pithampur in M.P.. Interestingly the campaign groups which were fighting to clean up the site in American courts, at the same time opposed the High Court ruling that the Government of India should clean it up.

Concerns about the role of foreign funding

Activists and NGOs who in the early days came to support the movement led by *Morcha* gradually withdrew and formed their independent groups. Many agencies and networks emerged to support Bhopal in Delhi, Mumbai and in other countries, and a huge international funding effort is channelled toward these ends. This process was started after a meeting of international funding agencies and NGOs on November 25th 1985 in London and concluded with the decision to create an international funding network for Bhopal and activist groups working for Narmada valley in M.P.. This helped *Narmada Bachao Andolan* (NBA - Save Narmada River Movement) to come into existence in 1987. I am concerned that some of the Indian organisations in this network may be connected with Naxalite movements in M.P., A.P., Orissa and Maharastra. If so, it is possible that international funds collected in the name of Bhopal had been diverted to such other organisations. It is a subject to be investigated by concerned agencies. The involvement of huge amounts of foreign funds has shifted the priority of the movement towards fund raising, leading to extremism, exaggeration of situation, production of false data etc. This has been become a characteristic of NGOs for getting more funds. Interest has shifted from seeking solutions to problems, to exacerbating conflict and creating deadlock on such decisions which are really in the interest of the victims.

In conclusion

Any social issue like Bhopal gas tragedy attracts a lot of activists nationally and internationally. If these people are there to do their research then that is okay but if they have come to help the people and work to change the situation then they should not hesitate to participate in their problems. If the outsiders come with the intention of helping, they should not back out when the occasion requires their intervention. When we came in the victims looked up to us for leadership because we had created that hope among them, so it would only be unfair to shirk that responsibility.

Outsiders and short term visitors have their limitations and cannot participate in the movement perpetually. People at the grassroot level have to take the lead. Outsiders can only help them to create their own instrument of struggle, the People's Committees, because they are always responsible for taking their struggle forward in the long run. Only such people could identify and offer solutions to the local problems. We link up the grassroots struggle with a wider perspective of movement. This is not something extraordinary; it is how things have worked everywhere in the history of the struggle of human civilisation.

The movement is constantly in conflict with the Government. We would fight against the Government but we would also use it and support it when necessary in the people's interest. I think it is important to recognise that we do not have any ancestral dispute against the Government, which consists of elected representatives. It does not make sense to oppose it all the time unnecessarily as some groups, especially foreign funded NGOs have always tended to do for so-called ideological reasons.

Rabiya Bee

Formerly *Bhopal Gas Peedit Mahila Udyog Sangathan*
Bhopal Gas Affect Women Workers' Union

Rabiya Bee had no history of political activism before she started work at the economic rehabilitation work sheds. She emerged as a fearless and articulate leader, formed and became first chairperson of the women workers' union *Bhopal Gas Peedit Mahila Udyog Sangathan* until family needs took her away from the front line. She remains a powerful advocate on behalf of survivors.

The whole system at the centre was corrupt

After the gas leak, Nirmala Buch, who was the wife of a government bureaucrat, started an organisation named *Swavalamban* to generate employment for poor women and widows of the gas disaster. The centre provided stitching, knitting, embroidery and jute work to the women; this ran for one and a half years. There were around 300 women in this centre and around 75 staff members who were divided into various departments. I was in the cutting department with 12 other women. I was good at my work because I had prior training from my father who was a popular tailor in Bhopal. My father used to train me in a lot of techniques. I would teach the other women these techniques, like making paper cut outs of the measurements, which helped them improve their efficiency and output. We would cut the cloth for around 300 frocks per day as compared to the 100 we did earlier. This increased our wages from Rs.40 to Rs.80 per day. This is when the exploitation started and Nirmala Buch began giving us limited work so that we would make less money.

The whole system at the centre was corrupt. People at the top made money everywhere, they got commissions at every level. There was a chaos when tenders for goods were opened; people would find ways to get commission on the smallest things like buttons. So I devised a strategy for the women to make some extra money as well – I taught them techniques to save cloth scrap in a way that they could use it to make some extra money. There were a lot of scams: big spools of cloth for centres would be stolen in transit. When everybody at the top was making money why couldn't we make some money? When the manager interfered I threatened him because everyone in the system was making money and he had no right to stop us, it was our right. I made sure that I took only as much as the other ladies, I did not take privilege of my position.

They were yet to taste the real power of women

When I started working I did not know what a Chief Minister was. I was poor, looking for a job, I passed the stitching test and later I applied to the cutting department and got through that as well. I was 28 years old at that time and I had 5 daughters. We did not know what a union could do or what it was. When Nirmala Buch began exploiting us it would make me very angry but I somehow continued to work despite the exploitation because I had a small baby to feed. Soon I raised objections and then they pointed me out to the other women who did not object to this just to isolate me. So I began talking to these women to motivate them to join me. The women slowly began to get my point and we spoke about this more regularly at lunch/break time.

Then ideas to make this group stronger were proposed in order to build pressure on Nirmala. A proposal to stop the cutting for a day was presented in one of the conversations and it was accepted because that way the centre would come to a standstill and work to all 300 women would stop. When women began to raise questions, the supervisor of the shed Usha Tai brought this to the notice of Nirmala Buch. She complained that I had organised all the ladies in the cutting unit.

Nirmala came up to me and enquired about my background and she found out that I was there because I had applied and passed the tests. I was not a widow which was one of the qualifications required to work at the centre but I was abandoned by my husband. She threatened to fire me but I warned her that if she did so all the 23 women in the cutting department would also stop working. The women supported me.

A senior woman from the group had a husband, Majeed Khan, who was working in PHQ (police head quarters) and he supported my views. Then we began getting ideas, the first one was to go to the Chief Minister but we had no idea how to approach him, we had no petition, no banner, nothing. We still went ahead with the plans, we reached the CM's residence and met the security guards who did not permit us to enter the premises. We insisted, so he asked what we were there for and he explained the whole concept of a CM to us. He also explained to us the concept of the union and advised us to form a union.

We took all this information back to Majeed Khan and he told us to choose the leadership. They all chose me as the president because I had the oratory skills, it was Gods' gift, and I could speak effectively. Vice President was Munni Bai Dubey, Sarita Mishra was elected the secretary and Rehana Begum was the treasurer. After this selection was done, Majeed Khan accompanied us to Indore for registration. We needed money for that so we went to all the shops in Itwara, Mangalwara and Patra where all the heavy machinery stores were located and begged for donations.

Then we stopped all work. When Nirmala learnt about this she shut the shed and there was a lock out at the Swavalamban centre for a week. Then the women from the *silai* [sewing] unit also joined us because they were anxious to know what had led to the lock out. Some supported us and some opposed us but we went ahead with our plans and registered the union.

Our first meeting was at the Central Library near the Shajahani Park, around 300 women from the entire centre participated. Then we took our first rally to the CM's residence. We were underestimated at that time by the Government but they were yet to taste the real power of women.

The movement handed the centre over to the state

Abdul Jabbar Khan was brought in by a woman member named Nusrat Jehan because she felt that we could use some assistance especially with the writing and clerical work. We also had educated girls with us, my niece Shajahan Bee was a graduate, but she did not have any experience in the field of social work to be able to write petitions/applications/pamphlets. We did not want a secretary, which was what most of the educated girls were capable of doing, we needed someone who could think and advise us. Further we were all women, we had social restrictions so we preferred a man who would be like a brother or a son to all of us. It is true that women led the movement but we needed a man, it is very important in life, there needs to be a balance, the way God made it. We had a meeting with him, he understood my mind and offered to be our advisor.

I used to live at Chowki Imambada, I gave Jabbar a place in my house and our work started off very well. With his assistance we began our sustained campaign against Nirmala Buch and she began feeling threatened. There were attempts on my life, she tried to break our union in every possible way. Nothing worked because all of us had a unique motivation to fight; most of us were also mothers at that time so we were also looking at securing our future.

Thus the movement started and we took the centre away from Nirmala Buch and handed it over to the State Industries Dept. Then more and more women began joining us, our membership grew tremendously.

We started to take up other issues: civic issues like cleanliness of the streets or medical care at hospitals: we would organise demonstrations on these issues. These were primarily introduced by Jabbar and Alok. They would tell us what to do and who to target. I would ask why we were taking on these other issues and their reasoning was to get more people on our side by raising their issues as well as ours. This tactic was beyond me because I felt like we were opposing the Government just to gain popularity, it was not right. That was the reason for the rift between Jabbar and myself. The main issue was gas relief and what we were demanding from America. When the money was given to the Indian government it should have been distributed immediately. There is also no reason to inquire into what the money would be used for or what the victims did with it. The money belonged to the victims and their families, if they purchased luxury goods with it nobody should have any problem because they had lost their loved ones and they could use the money for anything they wanted.

No better religion than Islam and nobody worse than a Muslim

I joined the organisation in the *burkha* because my husband was a Maulvi (a muslim priest) and I was from a very orthodox family. I gave it up because in our society women in *burkhas* are looked down upon, they are considered stupid, illiterate and without manners. When we went to government offices wearing our *burkhas* officers would treat us very rudely and a woman in a sari would be treated with respect

It is not communal, it is social conditioning which I cannot explain or blame on anyone in particular. It is a sort of discrimination against Muslims. I feel there is no better religion than Islam and there is nobody worse than a Muslim. The Muslim people have made their own religion the worse because of their actions. The *burkha* got a bad name because people did bad things in the *burkha*. I have personally observed the difference before and after I gave up the *burkha*.

If a woman cannot improve the life of her family, she cannot improve society

I had to leave the *sangathan* in 1990 because my husband was diagnosed with cancer of the throat. That's when the American tour was organised and I asked Rehana to go instead of me. I had five little girls and one young boy to look after so I could not go. This was also the turning point in my career as a leader because I came face to face with the politics of social activism. There are a lot of social stereotypes about women social activists. The word *neta* (leader) is more like a swear word. My husband's health was critical and if something were to happen to him while I was away my children would have never forgiven me. So I resigned from the *sangathan* duties and Rehana took my place. This was also the time when Rehana and Jabbar were very close. Sathyu was also a supporter. I was confident that the fire that I had ignited would burn for years to come even long after I am gone.

My husband died in 1990 and within two hours of his death my father also died and that disturbed my mind a lot. Within a year of these tragedies my mother died and six months later my young sister died leaving four of her children with me. So I was forced to shift my focus onto the family. I find all this other work useless. There was a time when I had the motivation and fire within me.

The disaster had long term repercussions. My daughter-in-law had gynaecological problems (ovarian cancer), a very common illness among gas victims. She was married to my son for seven years but we finally had to get them divorced because she could not conceive and the doctor warned that my son would contract cancer too. So I quit largely because of my family and the duties I had towards it. A woman's first priority should be her family; if she cannot improve the life of her family and children she cannot improve the society.

Rehana married Jabbar, Sathyu started his own organisation, Rasheeda Bee had her own union. I was left alone. Jabbar did the same thing that Nirmala Buch did but just ten years later and he broke the organisation. I have my differences with Jabbar but I will give him due credit: the truth is that Jabbar introduced us to our true potential, we had it in us but he brought it out. I would not ignore that. He taught us how to think differently and brought us away from our rolling pins and dough kneaders. He taught women how to fight for their rights.

An ego will be brought down by a small person

My friends still ask me to join the movement again but I cannot lie or deceive people so I will not last in the movement in its current state. When we started we were innocent, we did not know how this can be used to make money or gain power. Six years down the line we realised that this was a pimp market. I learnt that an illiterate person becomes a bigger threat than an educated person in such situations. The educated think that they can use their intellect to fool someone but if that person gets an ego about it then he will be brought down by a small person.

My message to young people who are into this work is that they should do it as long as they can do it sincerely. If they lose interest they should quit. People who they claim to work for can do without them, they do not need their help or they do not insist on getting help from social activists. People can survive with what they have. People who are not assisted by social activists also survive and people who know how to fight for their rights will do so without any assistance. So my message to the new generation is if they want to do social work they have to be honest and sincere, they should not take advantage or exploit. There is a lot of power in truth. And truth will also be your hindrance because it will cause a lot of problems for you and get you into trouble.

Mohini Devi

Bhopal Gas Peedit Mahila Udyog Sangathan
Bhopal Gas Affect Women Workers' Union

Mohini Devi joined the women workers' union *Bhopal Gas Peedit Mahila Udyog Sangathan* after its early establishment and soon became active in the leadership group. She is currently one of the most influential leaders and organisers in the union and its training organisation Swabhimaan Kendra.

In the struggle for economic rehabilitation, 2,300 women got jobs

Even before my involvement with the *silai* (sewing) centre or the gas leak I was active in my community. This started on a small scale but with time it increased and now I am often consulted and go out of the way to help. When relief was being distributed amongst the gas survivors, my *mohalla* (neighbourhood) was not given wheat and milk like the one across the drain was. I and my friend Lakshmi Srivastar, who also later joined the *sangathan*, gathered the women of the colony and took a rally to Chief Minister's house. My involvement in community issues has been there since before.

Some time in 1985 the Government provided us with relief, food and milk for about a year. They set up medical camps to give temporary relief to victims suffering from common problems of the eyes, stomach etc. They provided drops for the eyes and tablets to stop diarrhoea and vomiting. There was no organised program for rehabilitation. Food was distributed only after 'Operation Faith' on January 16. They started with milk (twice a day) then they started giving out food grains and later also introduced cooking oil (for six months). Grains were distributed for 1 year, milk for 8 to 9 months and oil for 5 to 6 months. And for economic rehabilitation they started silai centres.

I came to know about the *silai* centres when there was a survey from one house to the other, and we were asked to come and register our names. The women joining together happened here at the *silai* centres where they met and discussed and issues came up. For example, we realised that since only 300 women were given employment, this was too small a number compared to those who needed it. And during this struggle for economic rehabilitation, 2,300 women in total got jobs at the *silai* centre. Earlier there were a few *silai* centres run by Nirmala Buch, but they were closed down. The women brought this up and protested, and they were later reopened. But this time they were opened in larger numbers and 2,300 women were given jobs. Soon the women realised that they needed proper health care and medical facilities to work and to live. So one of the first major issues that we picked up on beyond the *silai* centre was health care. We did this by protesting and rallying, and also writing letters to the government. And it is because of this that today Bhopal has hospitals like Bhopal Memorial Hospital or Jawahar Lal Nehru Hospital.

The authorities used to get terrified of us protesting

When women got together at the *silai* centre we realised just how many atrocities we were facing that we should speak out against. One of the first things that brought us together was the first time the authorities tried to close down the *silai* centre. Later, the second time they tried to close us down, Abdul Jabbar also came into the group and helped give us direction. He taught us a lot about the importance of struggle and the fight for justice. The women decided to conduct open meetings so that everyone could gather and talk about the issues that affected us. It started with just a few women and then news spread and more and more people gathered and our strength grew. In the early days meetings would be held every day since women would come to the *silai* centres daily. Later it changed to 2 days a week and now we meet only once a week. But in the beginning there were about 4-5 women who got together: Rabiya, Tara Srivastav, Laksmi etc. All we did was speak about the issues and think about doing something about them. Later when Jabbar *bhai* came we found our feet. We knew about Jabbar since he worked amongst the people and was involved with gas relief work.

We had a working committee which would suggest issues which needed to be addressed, discuss which ones were most important and then start to work on ideas to carry out our campaign. To start with it was just workplace issues, and then other things started coming up. For instance, the lack of health care came up because people were missing work due to visits to the hospital and we recognised that you need to be healthy to do anything else. With time we understood things better and then people like Jabbar *bhai* and other educated people joined in who could guide us better and give us suggestions. The working committee didn't have any permanent members. Some have stayed, some have left and some are not involved any more. After discussing things in the working committee, the proposals would be announced in full meetings and with a show of hands we would see if people agreed with us or not.

My first demonstration was on 23rd January 1986, in front of the Chief Minister's house. However, the protest I remember most is the one in front of the Union Carbide office in Shyamla Hills on the 10th anniversary where people from all the campaign groups gathered in large numbers.

The authorities used to get terrified of us protesting, to the extent that close to 15th August (Independence day) we would all be locked up by the police and released later, just to make sure that any trouble was avoided.

Every problem relates all humans to the ones suffering

Issues picked up by the women were never restricted to workplace issues, they were open to the problems that people face over all. So their voices were raised for everything from medical health care, economic rehabilitation, compensation, environmental, social etc or for that matter the continuing rise in prices. For every problem, if you look at it on a larger level, there is a problem that relates all other humans not just the ones suffering in that place and time. This is why our solidarity went out to other campaigns also and likewise got the same back from them. Perhaps the most important fight that we are still fighting is to bring Warren Anderson back to India to face justice.

I have learnt the importance of the environment and the meaning of the term. Environment is what surrounds us and it is very important for a person to know what his/her environment is. The *sangathan* has also raised a lot of environmental issues like the misuse of the big and small lakes in Bhopal. The Hindus (for idol immersion) and Muslims (tazia) use it for their religious rites and we have raised the issue of pollution due to these activities many times. This does not mean that I am a disbeliever but this is of course not acceptable even for me as a person who believes in God. Similarly on a larger scale when forests are cut in the Himalayas its impact will be felt throughout India. I am not aware about the global environmental issues but I know that large dams are causing huge impacts on the environment.

We have also worked on communal issues. A lot of [reactionary] religious outfits tried to impose restrictions. Muslim women were abused and beaten because they wanted to step out of their homes and the Hindu women were criticised for joining an organisation led by a Muslim man. We also faced opposition from the Muslim religious leaders who were opposed to Jabbar *bhai's* views on the Purdha. I can say with pride that it was only after BGPMUS that the divide between Hindus and Muslims was bridged. People were very narrow minded before this and they would not interact with each other. This is what bothered the religious outfits. Even after the riots when we worked in the riot hit areas I went into Muslim *bastis* and I

Mohini Devi

fought with my Hindu neighbours who were opposed to Jabbar *bhai* entering our *basti*. My faith in the Almighty has increased because we have had so many victories. Personally I was in such a bad condition, almost on the verge of death and it is only because of my faith that I am alive and working with the sangathan today.

The issue of 36 wards was raised by BGPMUS when the interim relief was declared. The government only recognised 8–10 wards and it was only after BGPMUS took it to the Supreme Court that the 36 wards were included. The government then recognised 572,000 affected and 15,000 dead. We conducted surveys to prove our case and then filed a petition with the Supreme Court. We took some water and plant samples and tested them for contamination at labs in and outside Bhopal. Some scientists and doctors were also part of the organisation initially and they helped us with this and also gave reports to support the case. The Judges in the court ruled in our favour and the Government had to accept it.

When it came to the second compensation claim, the government officials scrutinised medical records of the victims and assigned categories (A, B or C) based on severity. We were called to the camp set up at the government hospital where they conducted a check up. They asked us basic questions like "are you sick" and took samples of our urine and x-rayed us. There was no logic, my husband did not undergo any treatment and he was given 'C' category and I was given category 'B' when I was admitted in the MIC ward for 3 months and underwent a 3 year course. So these categories were just randomly awarded on the discretion of the doctors and we had to trust the doctors. We also challenged this categorisation.

Every Gas Peedit group is raising its voice for some issue or another

There were around 15 gas victim organisations at one time and they all had their own agenda, they died out after their objectives were met and some just gave up. I feel that the differences between the campaign groups are not set in stone. The fact is that every *Gas Peedit* group is raising its voice for some issue or the other. The differences that arise come from their ideology and ways of working. Some may want to do things in one particular way and another may differ. The differences don't relate to the issues raised, the ways of fighting etc. The organisations have never united but they have worked together over issues or mutual relevance and then split up again. In 2001 Pension, Stationery and BGPMUS came together to do a survey, they all agreed to work on that issue, came together and split once the work was concluded.

When BGPMUS set up Swabhimaan Kendra we accepted government money for the training schemes. We will work with the Government when it is appropriate. This will not affect our fight because they are separate issues. We do not fight against the Government just on the issue of employment we have so many other issues. Just because the Government has paid us we will not stop fighting against it.

We also wish that our trainees will be provided with jobs later which could at least fetch them Rs 100 a day. Nearly Rs 70 *crores* were spent on making the work sheds that promised jobs to 10,000 gas victims and today they are useless.

Abdul Jabbar Khan

Bhopal Gas Peedit Mahila Udyog Sangathan
Bhopal Gas Affected Women Workers' Union

Abdul Jabbar Khan ran a small business before the gas leak and became secretary of a community committee in the early days after the disaster. He became leader of the women workers' trade union *Bhopal Gas Peedit Mahila Udyog Sangathan*, a position he still holds as well as running employment skills training NGO *Swabhimaan Kendra.* He is an influential leader and a key public face of Bhopal survivors in India.

On the 23rd January 1986, on the anniversary of Subhash Chandra Bose, the *Bhopal Gas Peedit Mahila Udyog Sangathan* (BGPMUS) was founded at the economic rehabilitation work sheds for gas affected women. Twelve months after the worksheds opened, an attempt was made to close them down. The workers formed a union and since some of them knew me and knew that I had some experience in social activism, they asked me to convene it. I was 28 years old. The first demands were simply to keep the work sheds open and to increase the amount of work for the 300 employees. I argued that the union should have more ambitious demands: we should also get economic rehabilitation for all those who were in need of it, men and women. In July 1987 we succeeded in getting work for more than 2,300. Then in the same year, State Industries Minister J. Vengal Rao inaugurated a big industrial estate in Govindpura to provide 10,000 direct and 10,000 indirect jobs and in a way this too was a result of our movement. Our demand now is that all the 152 units in the estate should be operational and only gas victims should be employed there.

Later the union took up medical care, compensation, environmental concerns etc. Initially the members were unhappy with these new issues and used to question why we were focusing on them, but their ideas changed. Now they also raise wider issues such as women's rights or Narmada Bachao Andolan or communal issues. These were women who were not politically aware. At one time I did not have that kind of political awareness either, I was a *mohalla* (neighbourhood committee) volunteer in *Zehreeli Gas Kand Morcha* and I became the General Secretary of Rajendra Nagar committee, in charge of improving conditions in that area. My political awareness came from experience, so I have always tried to raise political awareness amongst members in the *sangathan* so that larger issues can also be understood and addressed.

I have lived in Bhopal all my life. My father worked in the Bhopal textile mill. He was also sent to the World War as part of the INA. He was a freedom fighter and he lost everything in the partition after which he moved to Kanpur to work at the Victoria Mill Company. My father moved to Bhopal in 1958 and we have been in Bhopal since then. He died in 1986 because of gas exposure.

Social work was in my blood, I used to help a lot of people in my neighbourhood, with things that their own children could not do, like medical care for the old people. I used to take them to the hospital on my cycle and sometimes even on my shoulders. This was also in a way because I had experienced extreme poverty and I knew what it was like to go without food or medical care. I could relate to these problems because I had been through that experience. I had faced similar hardship in my education. I have worked in a brick kiln, pushed hand carts when I went to school. I started working at the age of 12 years and I have not been to college. I got a job at

Tata Exports in Dewas. Then I started a tube well boring business in 1981 which fetched me around Rs.5,000 per week. After the gas leak I abandoned all my business and started working here. I have fibrosis in both lungs as a result of gas exposure and have committed myself to the struggle of all gas victims.

Immediately after the gas relief, when the *Zehreeli Gas Kand Morcha* was formed, I was the only one in the group who was from a poor, less educated background. Most of the people at that time were from a rich and educated background and they used to look down on me. But I could identify with the issues which people were facing more than they could. Because of my background, when a person told me that he had not eaten for 2 days I could relate to it.

So in the *sangathan*, I feel that we somehow could relate to the everyday problems and hardships of the victims, we touched those problems in a way that the others could not. The others looked for ways to give issues political twists and attract national and international attention. Issues of prosecuting UCC and Anderson were on top of the agenda for them: in that scheme of things, the issues that we raised, about employment, rations, medicines, they were dwarfed or incongruous. Their canvas was large and these issues were small on that canvas. But these were the issues that brought our organisation success from 1986 to 2000. It did not mean that we did not raise other larger issues but these small issues were a priority for us.

We also succeeded in a lot of these people's issues. We got the Rs.200 relief from the Supreme Court on 13th March 1990, we got Rs.950 per month pension for widows through an order dated 4th May 1989 and through the same order they further won Rs.1,000 and Rs.3,000. On 28th May 1993 they won a relief of Rs.2,000 for 3 years. It is important in a movement to have your primary focus on what the people want, which in the case of the gas survivors is work and compensation. Then if you win a few victories in the things which are important to them, then you can fight on the bigger political importance. Otherwise you can lose the people's trust. If you work in the public field it is extremely important to gain and maintain the trust of the people. You have to make everything public, even your mistakes, and that's what we have always done in BGPMUS.

Some people have questioned how it is possible for a man to lead a women's union. In the first place, the women who formed the union were illiterate and had no experience of running a movement. Many were behind a *burkha*, they feared the cops, they feared their men, they feared to travel outside Bhopal. Some of them knew me, they knew that I had a little education and experience of social work, and they asked if I would help them. Despite the name of the union, I always encouraged men to join too, and argued that we should work for women's and men's employment. Men turned out in huge numbers to fill out the compensation claim forms, and we completed nearly 101,000 forms. But I had a lot of respect for women and I also believe that women are more faithful to the issues than men. I knew from experience that women were more committed to their family and children, that commitment we saw in our mothers. My father was committed as well but I saw more pain in my mother towards the needs of her children. Women are much closer to household problems than men, so they are more motivated. It's also my experience that if women begin to believe in something it grows stronger as they make progress. I cannot say the same about men; they try to take short cuts. I can say confidently that if the population of men in this organisation was more it would never have come so far.

For the first 10 years of the movement it seemed like a good idea to involve intellectuals just as they were active in the NBA. Now such people think very lowly of the Bhopal gas movement, they think it's a nuisance. They never have it in them to struggle. I feel that they could not connect to the problems of the common man because their experience was all book based. Their activism was jargonised but we would also go and fight with the local ration shop owner if four poor people did not get their ration. They focussed more on policy level intervention, we believed in that too but not at the cost of ignoring the four poor people who need our help.

During the British rule most of the intellectuals were in important positions in the system and they were the main hindrance to the freedom movement. It has been the same with the French revolution and the Russian revolution. The intellectuals are always with the rulers. So I would say that the uneducated people who do not possess 'literary' knowledge are the ones who can bring justice, much more than the educated.

The middle and the upper middle class live in a world of their own. Arera Colony, Shahpura, Idgah Hills: it's the same throughout the country. They have their own schools, hospitals etc. so they do not empathise with common problems. They take on issues only superficially or if they are fashionable. It is like *Diwali* festival, the poor are only remembered on occasions. You need to have the pain and commitment in your heart to work with the poor. About 5 to 7% of the gas affected population are rich and of that number, how many people have actually donated their compensation money? None! The poor will have to fight their own battle because the injustice has been done to them. I believe that all the major movements around the world have been led by the poor. The rich have only got in the way.

I was involved with the *Zehreeli Gas Kand Morcha* from within a month of its inception, along with all these educated people who formed it. They may not remember me joining but they later began noticing me because I had a fire within me. When the BGPMUS was formed, there were no serious differences between us and *Zehreeli* and in fact I had invited the leaders to join us, but they had a very different agenda. The main difference between the organisations at that time was that they were focusing on medical care whereas we were focusing on jobs. Our argument was that along with medical care, which they were stressing so much, people also needed jobs so they could feed themselves.

The four years from 1986 to 1989, were the best years of the movement. At its peak, BGPMUS was truly a mass movement. We used to have 10,000 or more women who had very little money for food who would make a point of participating in our meetings.

We would have huge numbers of people turning up for our events and we had demonstrations almost every day. The police would have to keep a constant vigil over us because they would never be certain when we would do what. There was opposition to us but we survived this mainly because of the faith people had in us, women more than men. Between 1986 and 1987 there were attempts to get our organisation banned on religious grounds because we were urging women to give up their *burkhas* which they said was un-Islamic.

On 18 August 1988 we filed a case in the Supreme Court demanding that government provide people either with a sustenance allowance or employment. On 3rd March 1989 an order was passed to start the distribution of rations again to the victims, which our organisation opposed. Our demand was that a sustenance allowance should be paid in cash and not a food ration. You see there was a lot of corruption with the distribution system. People had to queue up like beggars all day long. People had to miss work in order to get rations. It was humiliating so we demanded money so that people can choose the quality and quantity of food and provisions which they wanted to buy. Even then our first demand was employment. Sustenance allowance should be given only when it is not possible to provide employment. The discrepancies during the first relief were sorted out before the second relief. Some rich people were also receiving the relief so we demanded that income tax payers should not receive it. On the issue of widows' pension we had common sense that Rs.200 was not enough to run a household so we demanded more and won Rs.750.

On 22nd February 1989 after the settlement was announced, we ransacked the UCC office at Parliament Street in New Delhi. Then on 9th August that year we launched the Union Carbide Quit India campaign. A lot of national leaders participated in that - V.C Shukla, Arun Nehru, Sunderlal Patwa. We demonstrated at the UCC R&D Centre in Shyamla Hills. When the police ordered a *lathi* charge all the politicians (except Swami Agnivesh) disappeared and only the common man faced the violence. Once we were detained by the railway police for taking a rally on the train without tickets.

In India it is a citizen's right to travel by train for the purposes of lobbying politicians, but they tried to stop us. So we blocked the railway tracks until we were allowed to board the train that they had made us get off.

The *sangathan* is very committed to independence, especially when it comes to resources. When we started, members donated 50 *paise* and now after so many years we take only Rs.5 and that only from those who are able to pay. When an organisation takes money from outside, then it becomes dependent on others and this distorts what it does and how it does it. I am not against resources coming from abroad, but it is a big risk and we have decided to rely on our own resources. We have refused to work with Greenpeace in Bhopal for various reasons. We would have been quite happy if they had limited their involvement to technical and scientific expertise, and let the grassroots movement take the lead. But Greenpeace started to make statements on behalf of the movement with the intention of taking a lead. In a way Greenpeace used Bhopal to keep itself in the limelight: they like such international issues. If they honestly wanted to help they should have assisted local groups with science and expertise in preparing reports on scientific issues just like the People's Science Institute in Dehradun. They have made Bhopal into a commodity and destroyed its importance.

It is the same with the use of new technology. For those who know how to use it, it becomes a money making technique for themselves. They use it to prove that they are number one and people around the world are not made aware about the grassroots struggle. So some people misuse IT and people like us who cannot use the technology pay the price. It also affects tactics. The stunts which some groups get up to are very impressive and I support their aims, but it seems to me that they are mainly designed to getting international publicity. And in response to that publicity, the groups attract more money from abroad. International publicity has never been helpful for Bhopal. It is for the people who are directly involved in the publicity because they get the chance to travel the world. This has not provided one bit of support to the movement of the gas victims. All the publicity can only be accessed by a small percentage of gas victims because a majority of them are illiterate.

It is of course positive that US senators send a letter to our Prime Minister as they have recently done. That is very good but it would be even better if the senators wrote to their own government demanding the extradition of UCC and Dow because they are both registered in their country. Have they ever raised the issue of Warren Anderson's extradition in the senate? USA attacked Afganistan because their suspected terrorist who was responsible for September 11 was hiding there and they think it is not their responsibility to present Anderson who perpetrated the disaster? The senators can urge Dow and their government to do the right thing by cleaning up the factory and by providing the necessary relief and rehabilitation in Bhopal. They can urge the American people to boycott Dow shares in the market. There are so many things that can be done in the US and aren't being done. Instead they raise money for just a few campaigners.

BGPMUS is not a wealthy organisation but it is self-reliant. None of the people who work for the *sangathan* gets a wage. Those who are not working somewhere else manage from whatever their husbands or their sons earn. In times of need the *sangathan* helps them out, we all pool money, and we try to get them a job outside or with one of the training organisations associated with the *sangathan*. It is a fact that some people who come to organisations like ours lose interest very soon because we do not have so many resources or much money. People show initial interest but then disappear once they discover it is a regular job. But I believe that once people get an education it becomes impossible for them to work without money. Most of the people who will sacrifice their lives for a cause will be the less educated or the illiterate.

Now we have *Swabhimaan Kendra* affiliated to the *sangathan*, which provides training mostly for gas affected women, and a few men, in tailoring, embroidery, weaving, computer skills and so on. We get a little money from the Government for this training but we provide much more than the minimum level of training which the Government expects. So we are able directly to contribute to peoples' economic rehabilitation.

The main campaign for BGPMUS now is for 5-times more compensation. It is a matter of arithmetic that of the money which was allocated for compensation, only one fifth has been spent on gas victims. So all survivors have a right to the remainder of the money which amounts to five times what they have received. This is very important, people need money to survive, it is our right, not charity. Why were the victims of the September 11th 2001 given US$ 24 *crore* each? Is there so much difference in the value of American and Indian lives? Compensation is a very important issue to the lives of the victims and it has been very popular. Economic rehabilitation also attracted a lot of people, in fact the whole thing started with that. But it does not mean that we are not also demanding the extradition of Anderson or Dow Chemical taking liability. We have an ongoing medical care petition in the High Court; a petition for those who were left out during the compensation distribution; and we are also the interveners in the Dow chemical clean up case. In 1991 nearly 3,000 huts were to be demolished so we launched a movement against the "anti encroachment drive". We tried to stop the drive, when we did not succeed we took it to the SC and got a stay on the proposal. The poor in those huts still live there. I believe in fighting for the poor, the issues are irrelevant.

The movement as a whole has also worked very hard towards maintaining communal harmony in Bhopal. It has changed a lot of misconceptions and presumptions among communities because it brought people together and because people got together they realised that what the Hindus thought about the Muslims and vice versa was all wrong. The main reason for the communal divide was the ignorance among people. Nobody told the Muslim women to give up *burkha* directly but they gradually became aware and they understood the importance of the ideas. When they realised the importance of fighting and the *burkha* was a hindrance they had to give it up eventually. It was a natural process. They travelled to other cities which exposed them to different ideas which eventually led to a lot of awareness about a lot of issues. We removed the practice of untouchability from the union which was prevalent among both Hindus and Muslims. We did a lot of work during the communal riots which followed the destruction of the Babri Masjid in Ayodhya, and we presented a 325 page report to the commission.

Our country doesn't need development in the way that it is understood in the west. India has a population of 120 *crore* and there are 25 states so the western concept of development doesn't make any sense. We need a Gandhian model of localised development. When a city grows the Government proposes a dozen infrastructure schemes like fly-overs and roads but the root cause of the problem is bad development schemes at the village levels that forces people to move into urban areas. The model of development needs to be changed. We need to make our society and our villages self sufficient if we are serious about development. The reality is very conflicting. The products of globalisation are being imposed upon the rural population. Government is giving out free TV sets and opening new alcohol shops in villages, even in villages where the water supply has failed. This will only lead to destruction.

Yes, this is going to be a long struggle. There is a very popular adage of the common man: "the world survives on hope". We can all depend on that hope but at the same time we cannot give up what we are doing. You can never predict the timescale of a people's movement. The freedom movement lasted from 1857 to 1947, 90 years, sometimes you can start something but you cannot end it. I spent the most productive part of my life for the movement.

I do a lot of things now just because they make me feel satisfied with my life. I am no longer in the race for publicity and media, we solve so many issues every day that could get us so much publicity but we do not bother about it. We also try to resolve a lot of issues in people's personal lives. People have that faith in us so they prefer coming to me rather than going to some priest. I am careful when I advise them; my attempt is to patch things up to the extent possible rather than making matters worse by filing police complaints and stuff. These are very important for people to retain faith in the movement.

I have been hiding two things about me – my education and my birth, because I learned this important thing from my Guru Shankar Guha Niyogi. (Union leader and organiser of the Chhattisgarh people's movement, Shankar Guha Niyogi was assasinated in his sleep at the headquarters of the Chhattisgarh Mineworkers' Union in Bhilai in 1991). I strongly believe that all the major problems of the world have been created by the educated class. I also learnt from Niyogi to live among the people who you work with. I still live in the same labour colony house which I moved into at the age of 1.

Hamida Bee

Bhopal Gas Peedit Mahila Udyog Sangathan
Bhopal Gas Affected Women Workers' Union

Hamida Bee has become a key leader in women workers' union *Bhopal Gas Peedit Mahila Udyog Sangathan*. A fiery orator, she provides a significant mobilising role in the union's activism.

Widows have nobody to look after them except God and BGPMUS

The BGPMUS has been constantly struggling, fighting and also winning long legal battles. We have knocked on the doors of local Bhopal Courts, we won at the High Court, Jabalpur, and we won at the Supreme Court. We also fought against the state and central government. We have found such a wonderful brother in Abdul Jabbar to help us in our fight and we will fight unto death.

The gas victims have been suffering for so long with all sorts of serious cancers. The children of the survivors are sick themselves, so how will they look after themselves or their sick parents? Isn't this dangerous: people are getting cancers: uterus, breast, mouth, tumours?

My family is from Ibrahimpura. My mother had a lot of miscarriages and we were only two sisters who survived. I was married at the age of 11 and when I was 14 I gave birth to a son. Before the tragedy my husband was a driver and the family thrived on his income. His hand was paralysed after that night and his income stopped. I was skilled, I had trained in stitching and *zari* (embroidery) so I started this to sustain my family. I lost 35 people in my immediate and extended family to the gas. My husband lost 6 brothers, their wives, nephews. I lost my aunt, my mother, my mother-in-law, my daughter, two of her sisters-in-law, my daughters-in-law. We are still awaiting justice. This is a tragedy: the women who were widowed have nobody to look after them except God and BGPMUS.

Arjun Singh was the Chief Minister at the time of the disaster. He received the news of the gas leak. The chief ministers have the contact numbers of all police stations and all he had to do was mobilise them. Had he done that and sent out announcement vehicles, so many people would not have died. This is how I hold the government responsible. The CM and his family members were flown out of Bhopal instead of tending to the crisis at hand.

Another thing was the last rites of the victims. The Muslims were not given shrouds, women and girls were just laid in long ditches and buried. The Hindus were burnt in mass pyres of 400 to 500 people and the piles were set on fire with kerosene. Was the Government so heartless? This is our problem.

And why was Warren Anderson sent back to America by the government when he came to India, rather than putting him in court? We, the BGPMUS, have been demanding the extradition of Warren Anderson for the past 23 years but that has not happened. There is only one book of law and that had awarded a death sentence to Satwant Singh and Peher Singh, the murderers of Indira Gandhi. That was in this same India, so why is Warren Anderson still at large? Our real justice will be when Warren Anderson is punished. If the CBI (Indian Central Bureau of Investigation) cannot extradite Warren Anderson then we will continue to fight.

I used to give speeches wearing *purdha*

When I heard about an organisation that was fighting for the poor gas victims I started attending meetings near Ajab Ghar. I used to wear a *burkha*. I heard Jabbar *bhai* speak and he had a unique way of talking that touched my heart and I started coming for the meetings regularly and became more active. That's when BGPMUS was giving out forms to the public for membership and livelihood, people turned out in huge numbers.

I used to give my speeches wearing my *purdha* back then. During press conferences the reporters were concerned that I would be sick if I continued to wear this and even Jabbar *bhai* was concerned but I still did not give up. Then I met Shankar Guha Niyogi from Bhilai who was also fighting for labour rights. He was assassinated. He was a very good friend of the organisation so we all went there to meet his family and when we returned Jabbar *bhai* made me take my *burkha* off. I wore it again the next day but the reporters made me take it off. My perspective changed slowly, I felt very awkward for some time. Now I do not need to wear a *burkha* to show that I am dignified.

We used to raise all sorts of issues at our meetings concerning women's rights, like dowry, abuse etc. We have also supported movements like *Narmada Bachao Andolan* (NBA) and Shankar Guha Niyogi's Bhilai movement has received support from BGPMUS. We also supported the fishermen's rights movement in Bhopal. We worked with the NBA and even went to jail with them. After a meeting organised by the NBA in Maheshwar all of us were arrested: we were put in the old jail and the NBA activists were put in the new jail. We spent 11 days there and we demanded everything right from tea to soaps to bedding and we received everything because we bullied them while the people in the new jail received nothing.

Cows, goats, chicken were compensated for but not a child

The government has not helped us a bit, they are not even talking about the Below Poverty Line cards for the victims. The victims are living in rented houses and they pay all their savings towards rent plus they have to pay more for their provisions. They have not listened to the problems of the widows of the disaster. The BGPMUS has been fighting through the law and it hopes that the law will certainly bring Warren Anderson to justice.

No political party has helped us: they have just watched. There are people in the parties who are gas victims but they don't care. Shankar Dayal Sharma who was the president of India is from Bhopal, Lakerapura which is my husband's home place. We expected him to do something since he is a gas victim himself but he never raised anything. The widows of the communal riots get Rs.1,000 per month. We worked with the victims of the communal riots as well. We arranged shelter and food for them.

First all claims were being accepted and then when the numbers were rising they passed a law that only victims above 18 years of age would be eligible. Domestic animals: cows, goats, chicken were all compensated for but if I have a child of 5 years old who has been exposed to the same gas, how come I cannot file a claim?

They also messed up the names when the claim forms were being filled in. Officials were called from Kerala and Madras who did not know Hindi. They put an extra line or removed a dot and messed up the names. Thousands of people could not claim because of these mistakes. We discovered this when Abdul Jabbar filed a case which went from the local court to the high court and then we finally appealed to the Supreme Court. Then while the case was going on, Arjun Singh went to America with the excuse of his wife's knee operation, along with Rajiv Gandhi and they reached the settlement.

We painted the court premises red and black

We challenged the settlement but the Supreme Court dismissed our claim. Prashant Bhushan called us and we began running around in a panic informing the public and overnight we gathered strength. The next day at least 15,000 people demonstrated in Delhi in front of the Supreme Court. This was 13 years back. When we reached there we ran amok with the judges and they ran. We painted the court premises red and black, demanding the withdrawal of the settlement.

Then we attacked the office of Union Carbide, around 15 of us. We ransacked it. We came down and burnt an effigy, the police came and we faced the police. Then we held a press conference which was attended by 400 journalists.

Once the settlement had been announced and the V.P. Singh government came in, we demanded disbursal of Rs. 200 relief from the Supreme Court. V.P Singh acknowledged the strength of the organisation and he invited Jabbar *bhai* to come and see him along with eleven of the women. V.P Singh enquired about our demands and Jabbar *bhai* told him that our application was dismissed by the Supreme Court so we want the Government to file an application and get a verdict out on this issue as soon as possible.

V.P Singh said: "courts take their own time, I can't guarantee that. What do you want from me?" So *bhai* said our case for compensation is pending in the court but the condition of the victims is getting from bad to worse so we want the Government to provide some financial relief. *Bhai* proposed Rs.500 but V.P. Singh declined because it was too much and they both agreed on Rs.200 per month. The government filed an application in the court and the verdict was in favour of BGPMUS.

This was just half the victory. Then we sat on a 13 day *dharna* outside the Supreme Court in order to have the victims below 18 years of age included in the relief. The court decided in favour of BGPMUS. The state government did not accept this so we started our *Jail Bahro Andolan* (fill up the jail movement). Then, during the 7th round of the campaign we had women from all age groups with us and we decided that we will not leave the jail. It was a successful campaign and we won claims.

We also fought for employment. When we got employment whilst the Congress was in power, 2,200 women were employed. The cloth cutting happened in J P Nagar. Each woman got 4 half pants and 4 frocks. It didn't help them a lot but it was a substantial support. They made 500, 700 or 1000 rupees depending on the time they could spend.

Then the government saw our strength increasing, we demonstrated all the time, we entered the Chief Minister's house and we would be in Delhi in huge numbers. So the BJP government tried to break the 2,200 women union. But we were too strong for them.

We will beat and straighten the system in government hospitals

We do not think that health care can be adequately provided by NGOs, and we will not take gas victims to *Sambhavna*. We will continue to beat and straighten the system in the government hospitals, we prefer to go back to them because they are established and run with the money that belongs to the gas victims.

As for the factory site, the principal responsibility for the clean up is the government that has ignored the issue for so long. Now if Dow has purchased UCC then it is also its responsibility to clean it up. However when clean up demanded by some organisations on the one hand, and then when it is proposed it is blocked by other organisations, I am angry with that attitude. I strongly feel that the government has to clean up because it has received so much help from UCC in the past 23 years. Dow came in later, first the government gave permission to such a dangerous factory and then it betrayed the people of Bhopal consistently.

We boycott chemicals in our daily life. We boycotted Coke, soaps, toothpaste, liquor and plastic bags. People use plastic bags to fill hot food which is very harmful when the poisons from the bag leak into the food. We have protested and shut down liquor stores and also held *dharnas* against lotteries, because they were spoiling family relationships. In short we will fight against any kind of injustice and we will be present in solidarity to causes that need our help.

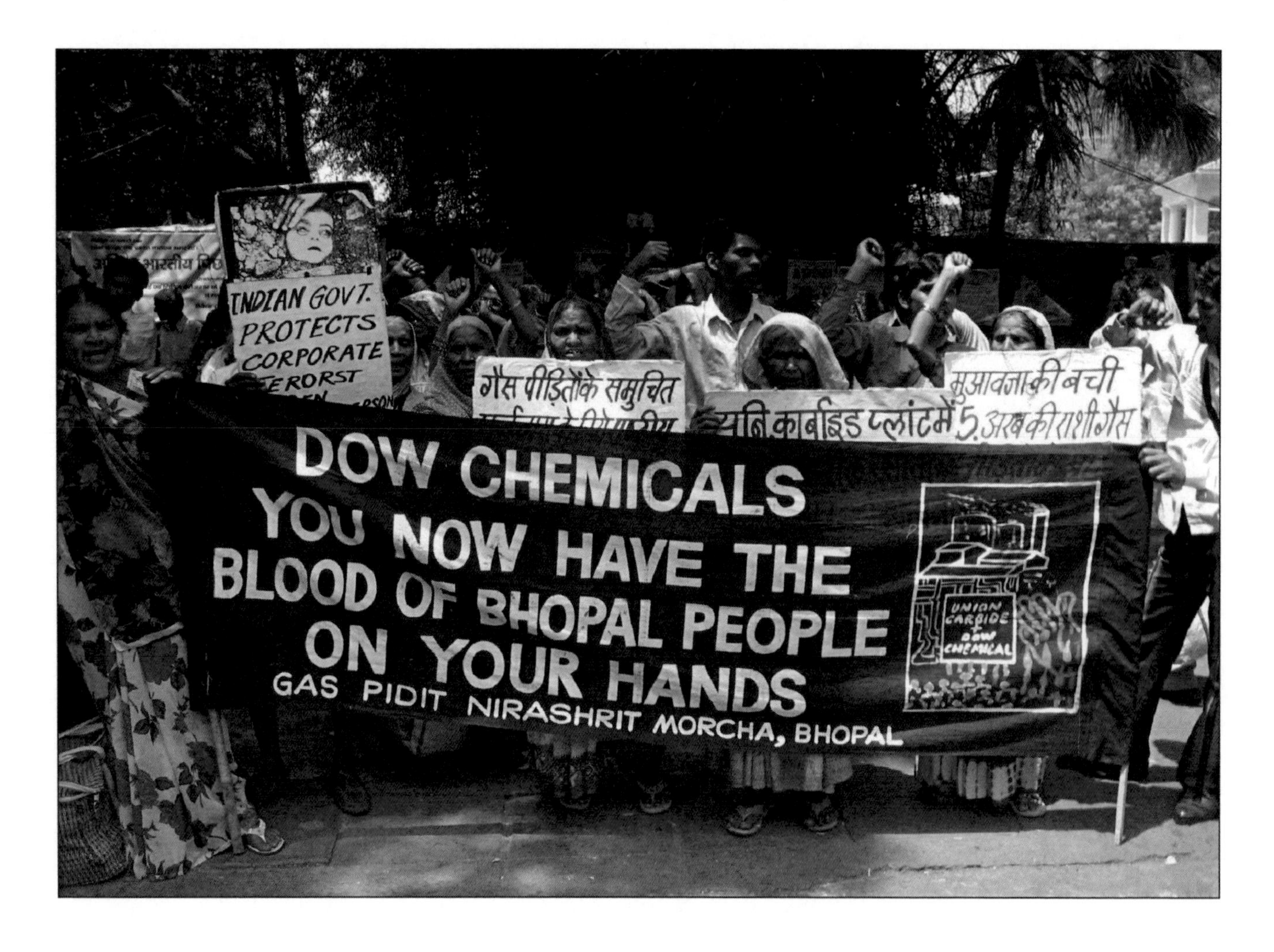
INDIAN GOVT.
PROTECTS
CORPORATE
गैस पीड़ितोंके समुचित
मुआवजाकी बची
अरबकी राशी गैस
DOW CHEMICALS
YOU NOW HAVE THE
BLOOD OF BHOPAL PEOPLE
ON YOUR HANDS
GAS PIDIT NIRASHRIT MORCHA, BHOPAL
UNION CARBIDE
DOW CHEMICAL

Rehana Begum

Formerly
Bhopal Gas Peedit Mahila Udyog Sangathan
Bhopal Gas Affected Women Workers' Union

Rehana Begum is an educated and articulate gas victim who was one of the founders of women workers' union *Bhopal Gas Peedit Mahila Udyog Sangathan*, amongst the early leadership and a later chairperson. She was married to Abdul Jabbar for fourteen years and left union activism when the marriage ended. She works in economic rehabilitation with the NGO sector.

A huge membership of 12,000 people, mostly women

When the women in the sewing centres first got organised, our union was called *Bhopal Gas Peedit Mahila Udyog Congress-I* [Bhopal Gas Affected Women Workers Congress-I [Indira]]. We had no idea about the concept of a union or about the concept of a state or what the Chief Minister was. We chose the name Congress-I because the symbol of the Congress party [the palm] was very prominent and we were familiar with the symbol. It was appealing so we picked that name and sign. There were no political affiliations. Rabiya Bee was the convenor and I was the treasurer. The name was changed from Congress to *Sangathan* only in 1988 as a result of Jabbar's insistence – we were not that bothered what it was called!

I was married to Jabbar in 1985. It was my second marriage because my first husband had passed away before the gas leak. I was a widow with two children and Jabbar offered to do *Nikah* [a second marriage in Islam]. I was married to him for 14 years. Jabbar was brought into the *sangathan* by a lady called Nusrat who was also a member of the union. She and Jabbar were neighbours and she introduced him to us. We were also looking for some good advice at that time so our convenor Rabiya Bee liked the suggestion. When Rabiya Bee met Jabbar she told him that they were looking for an advisor but he could not be a part of the organisation because it was a woman's organisation. However, Jabbar joined four months after the formation of the union and started acting like its leader and that was when the mistakes started to be made I now realise. It wasn't because of his education: I am more educated than he is. It was only because we were mostly homely Muslim women who had no experience of negotiating or campaigning and frankly did not have a clue about anything. So Rabiya Bee felt that a man could be of some use.

We used to conduct surveys in the *bastis* but they were more for membership and used to recruit people into our union. We went around explaining our work to people and got them to fill out membership forms and a small contribution. We had a huge membership of around 10,000 to 12,000 people between 1987 and 1989, mostly women.

We wrote our demands on a *dupatta*

The way that the *silai* sewing centre operated was that we would each deposit a sum of Rs.50 and we received these receipts which had individual numbers on them. These numbers were also put inside each set of clothes that we delivered. If damaged or improper goods were identified then they would be returned to the women to whom they belonged. All this was entered in a register that was maintained by the centre and entries were made every day.

When the sewing centre was closed down, we were left without money or jobs, we had no other choice. So that's when a few older women suggested a rally because they remembered seeing political parties do it during elections. Our first demonstration was at the Chief Minister's house. We did not have anything at that time so we broke some tree branches for the banner and wrote our demands on a *dupatta* (scarf). We didn't even know the way to the CM's house so we had to ask for directions.

After 1986 when we succeeded in securing our jobs, people from outside, politicians and trade union leaders, they met us and advised us to demand more compensation. That's when the issue of compensation was introduced. After that we began mobilising more women and that's how our strength grew and we began demonstrations in Delhi.

All that has been achieved over the past years has been the work of the women and Jabbar has always been in the background. But when women began making progress without him I suppose he felt threatened and his male ego was hurt. Jabbar's behaviour towards the women in the organisation changed. He started bringing in new people who could support him and sideline us. All of us began distancing ourselves from him but not from the organisation. Some women quit whereas others were determined to stay because it was the women who had done the work and had made sacrifices to build the union. The women retain all the registration papers and Jabbar continues to be an advisor. I have not quit, I have just taken a break from the organisation because of my personal financial situation and my personal problems with Jabbar. My problems with him began when I was selected as a trainer in the jute bag making unit run by the government. When I was promoted to the post of a teacher he felt very threatened and it led to us splitting up. I took leave of the organisation in 1998.

Employment is the most important thing

The main mistakes which I think Jabbar has made are to move away from our core demands of employment, and to change the way that the union was funded. Employment is fundamental and the union should have concentrated on this alone. I have seen extreme poverty and I know how much difference employment can make to a person's life. Without employment a person cannot have access to medical care, food, housing, clothes, it is an important issue. So rather than campaigning for health care or environment, we should just focus on employment. Jabbar began introducing the other issues of compensation and medical care. A person can access good medical care if he has a good job, but good medical care alone will never ensure prosperity. Even to access the poor medical facility at the government hospitals a person either needs money or contacts, a poor person with none of these will have to give up on life. We made the *sangathan* to provide employment to the gas victims and to assist them in accessing medical care.

Employment is the most important thing which the gas victims need. I have been to Sambhavna clinic once and I felt that it was very extravagant, setting up something of that style in a poor community is inappropriate. It would have made more sense if all that money had been used to provide employment to people. Rasheeda Bee should have done the same thing instead of making that institute for disabled children [Chingari Trust]. The money that she received from the award could have been diverted to generate employment for young men and women.

When we started, all the funding for the union was raised by local donations which were collected by the women. The money was spent on stationery, letter pads, banners etc. A demonstration would cost about Rs.1,000. We used to collect this money wearing our *burkhas* which was a very big deal and a real social challenge. We never accepted any foreign funds or funds from the government. Later we began collecting Rs.5 each from the 2,378 women members of the union. The demonstrations in Delhi were sponsored by people: food and stay would be arranged, N.D. Jayprakash and Deena Dayalan from The Other Media were our supporters back then and they would make the local arrangements. Deena's organisation was run with foreign funds but he would never hide that, it was public. We used to charge the members Rs.10 per ID card during these rallies and that money was spent on food and transport. We never accepted any donations from political parties. All that has started after Jabbar Khan came into the picture. He had allied with the Congress leaders before and now he has friends in the BJP party.

Sathyu was also a supporter of the union but he and Jabbar had some difference over foreign funding. I do not see that as an issue because it is a personal choice, eventually it has to reflect in the work each organisation has done. Sathyu was the one responsible for making Bhopal an international issue. He had a lot of connections abroad and I made my foreign trip only because of him and Ward Moorehouse.

When I went for my foreign tour in 1994 I lived with families instead of at hotels, because Ward had arranged for that. But when N.D. Jayprakash [of BGPSSS] gave a budget for reimbursement he arrived at a figure of Rs.3 *lakh* when I had not spent that much money. I objected to this. Sathyu and Jabbar operate similarly with foreign funds.

One day I will get involved in campaigning again

Now the *sangathan* accepts money from the government. The *Swabhimaan Kendra* is the training scheme set up by Jabbar under the *sangathan*. I work at the state emporium head office and my office works with the *sangathan*. The women who have been selected for training are supposed to receive Rs.1,500 from under the government scheme. Goods from the *Swabhimaan Kendra* go to the head office to be inspected and the bad goods are returned to them. All these are rough goods that are made by women who are being trained so it is not very good quality. We do not make any extra payment in exchange for the goods; the costs are all covered under the Rs.1,500 that the government pays per trainee to the *Swabhimaan Kendra*. But the organisation only pays them Rs.500 hence making a commission of Rs.1,000 per trainee.

The *Rajeev Gandhi Memorial Gas Victims Rehabilitation Centre* run by Alok Pratap Singh is an independent NGO that is registered with our head office. The work is proposed by the NGO and if it is acceptable to our department the work starts. All the goods produced at such units are displayed at exhibitions, 10% of the proceeds go to the communities.

There is still a lot to be done. One day I will get involved in campaigning again and will start a new organisation. I will work on the issues of employment and the abuse of women. I will name it *Mahila Udyog Sangathan* because we still retain the registration rights.

Syed M. Irfan

Bhopal Gas Peedit Mahila Purush Sangharsh Morcha

Struggle of Bhopal Gas Affected Women and Men

Syed Irfan is leader of *Bhopal Gas Peedit Mahila Purush Sangharsh Morcha*, an organisation which he formed when he left BGPMUS. He plays a significant role within the International Campaign for Justice in Bhopal as organiser and mobiliser as well as providing individual casework support.

Zamindars, trades unions and peace marches

My parents were *Zamindar*. The *nawab* of Bhopal had given my forefathers 885 acres land near Raisen. When India got independence in 1947, a new law abolished the *Zamindari* system and our *jagirs* (land) were taken away from us. Whatever land was left with us was tilled by my parents but by the time we grew up, we had very little land. We were two brothers, elder uncle had two sons and the middle one had four. There was equal distribution of land after which each one got very little. At that time, I was in higher secondary. I passed my intermediate in 1963 and set up a newspaper shop near Raisen bus stand. In 1968, my marriage was fixed with the daughter of elder uncle. After one year of marriage, my newspaper business which was named *Solash News*

Agency was sold off to pay off the loans taken for my marriage. I came to Bhopal in 1971 and then in 1972, I was appointed as daily supervisor in the government milk scheme.

Then in '75, the emergency was declared and I was arrested along with my brother-in-law, who led the strike in the state buses. As a result of this I had to leave the job and got work with the national textile corporation. I was not getting enough work so I put up posters saying that I could not support myself and therefore I would commit suicide. So the general manager called a meeting with me and Yadav who was the local chairman of the INTUC (Indian National Trades Union Congress). It was through this that I was absorbed into the union which was affiliated to INTUC and from then on started to work with them.

When Indira Gandhi was assassinated in 1984, curfew was declared in Bhopal and the entire city was closed for many days. I participated in the peace march undertaken by the Congress and other welfare organisations. As a participant of these peace groups, I helped in monitoring the situation in many areas including the ones where Punjabis (Sikhs) resided. That December I came back from a union meeting about taxi stands with my brother-in-law and reached home around 12:00, had food and lay down. As soon as I lay down, I had a cough and there was smoke all around and a burning sensation in my eyes. I came out on the road which was just nearby and I saw a woman running in her petticoat and blouse with her child clutched by her side and men were walking in their underwear and *banian*. And the coughing did not stop. Then I said now we should leave. My wife got very worried and went to her mother's place in Bhopal along with the children. We went to Hanumanganj police station to file a report but there was no one there. We met no one. Then we lost all sense as to where we were, such was our state of mind and I don't remember whether we had fallen down or what. Someone caught us by the hand and put us in a truck coming from behind, they made us climb on board and dropped us at Misrod around 4-4:30 which is around 12-13 km away. By then, our eyes had become totally closed and we could open them only with great difficulty using our hands. Then the police came announcing that this gas had leaked from Union Carbide.

Issues were raised by the workers regarding safety precautions

During my work with INTUC I knew about the factory and its earlier mishaps – cattle dying and the loss of a worker's life - but the lawyer of Union Carbide India kept the matter quiet and paid some money to the worker's family and covered the whole thing up. People knew that the factory made pesticides used to spray on fields and other poisons to kill insects. In the earlier years I was aware of the issues being raised by the workers regarding safety precautions and health problems (I had relatives and friends there too). Also that when workers raised these issues they were ignored. After the gas leak some of the workers who could work were sent to other cities, others generally between ages of 30 to 40 were given jobs in different departments and the rest who were not capable were given some amount as compensation and sent off home. And the factory closed down.

When it was built in 1970 the areas around it were not as densely populated as they are today. But the population did increase with time and especially after plots were distributed by Arjun Singh, the Chief Minister. Most of the deaths were in and around the factory (J.P Nagar, Chola Road, Dwarka Nagar, Qazi Kamp) but also at the bus stand and the railway station - people didn't know about these as the bodies were disposed of into the Narmada River by the officials. Some say this was because of the lack of space and wood but others claim that it was to reduce the overall body count.

The government did provide some immediate relief by distributing money amongst the people on the 4th of December, but on 5th we read in the newspapers that officials had been handing out Rs.400 but making people sign for Rs. 4,000. So this was stopped the following day.

Neither the ruling party nor the opposition had raised questions about the gas leak

On 7th of December 1984 we filed a case against the officials of Union Carbide India and the supervisors in the factory. I and some other people from my neighbourhood and surrounding community formed a committee. This included railway colony, Phuta Makbara basti, Jahangirabad and Kabar Khana. We opened an office on the main Chola Road. Around the same time all these people came to our area and talked to us: Anil Sadgopal, Satinath Sarangi, Tapan Bose, Alok Pratap Singh, Ajay Singh, someone called Qureshi and some other people from Madras, Calcutta and Bombay came and helped. They met with our committee to ask for help and I would go to their meetings at their office at Chola Road. At this time, this group had no name for themselves.

After the gas leak our committee conducted a survey of all those people who died, the illness that followed and the losses in the nearby areas. We conducted the survey by dividing ourselves into groups and going from house to house and making a note of the deaths, illnesses and losses. This group consisted of both men and women and included Harlal Kushwaha, Harilal Patwa, Anita Verma, Vijay Rajwa, Janaki Prasad, Tulsi Ram Rajput who was an Advocate, Sayed Masiruddin aka Shahid. There was a lady who we used to call Sen Chachi (Auntie Sen), she was the *prachar mantri* 'propaganda minister' of the committee, since she was from the barber colony and in those days barbers were often sent to invite people for marriages and functions and they knew many people. The chairperson was Harilal Patwa and I was the vice chairperson. The *sangathan* was called *Bhopal Gas Peedit Sangharsh Morcha*. We formed the *sangathan* because neither the ruling party nor the opposition had raised questions or spoken up about the gas leak. Only the Communist Party raised their voice but it had very little strength, and this went quiet after the death of Shakir Ali Sahib.

I still retained my links with INTUC in the following few years but my work was more towards '*Kaumi Ektaa*' – 'religious unity' since there was increasing communal tensions and unrest between the religious communities.

There has been no other rally as big and as powerful as this

By 12th January 1985 the group of intellectuals in Bhopal had come up with the name *Zahreeli Gas Kand Sangharsh Morcha*. They decided that they, along with our committee, would gather people together and walk to the Chief Minister's house via UCIL. About 15,000 people protested that day and it was a strong and vociferous rally. There has been no other rally as big and as powerful as this one to date. Fifty six people were called in to meet the Chief Minister as a part of the delegation. After this rally the people of Bhopal were provided with rations and also promised a survey by the students of TATA Institute of Social Science, which later formed the basis of the interim payment.

Later when some of the demands were not met, like medical health care and compensation, they demonstrated in front of the CM's house for 8- 10 days.

Next thing it was decided to start the *Rail Roko Andolan* (Stop the Rails Movement) on the 12th February 1985. But by then Arjun Singh had been able to weaken the movement by buying off some of the members of the ZGKSM. There were differences amongst individuals about the protest idea and how they should work and soon they went their separate ways. But the immediate affect was that some went against the idea of *Rail Roko Andolan* and informed the police and the plan did not succeed.

***Lok adalat* is a farce**

Then later when there were divisions, a group of ladies formed the *Mahila Udyog Sangathan* and asked Jabbar to lead them. He could speak very well and used to live decently. Neatly clad always and he was very bold. He was neither scared of the police nor the government. He used to curse the government openly and probably, seeing his boldness, the women thought of making him a leader who would do good work for them. They did not have any office so meetings took place behind Radha talkies, where the sewing centre was, or under the bridge of Bharat talkies. There were other groups at that time. I remember there was also *Jagriti*

Mahila Sangh under the leadership of Rajni Telang from Congress and a Christian woman.

Lok Adalat (people's council) was set up at that time. Our organisation was in existence but we could hold only 1 or 2 programs in a year. We had stopped holding neighbourhood meetings of *Bhopal Gas Peedit Sangharsh Morcha* and when *Lok Adalat* was set up we attended the meetings. We were called to the *Lok Adalat* but then we saw that they were neither listening to the victims nor doing anything for them, even though we tried telling them. So we opposed it and walked out shouting slogans. The police chased us and we ended up at Shahjahani park where Jabbar and the women were standing in opposition but quietly. When we came shouting "*Lok adalat* is a farce" they immediately supported us and Jabbar *bhai* called us. So all of us joined them in their meetings. We used to stand along with them every Saturday. When they used to hold meetings, the workers there were our friends so we used to stand along with them. Their strength was about 3,200 women. We worked for 9-10 years with them.

We separated in 1999. Jabbar said a few things which seemed as bragging that only he has the brains in the movement. I am a little worldly too. I have run unions and worked in different places and as a result, I have had to tolerate quite a few things. There is one very wrong trait in him, we tried a lot and joined hands with Namdeo's party with Rashida Bee's party, stationery labour union. None stayed for more than two years. We were not allowed to put up our banner alongside theirs. They consider the *Mahila Udyog Sangathan* the only organisation having empathy for gas victims and therefore, the only organisation whose name should be projected. That is why we separated from Jabbar *bhai* because Jabbar *bhai* talked and listened to only himself. Many journalists, political leaders, even ministers told Jabbar *bhai* that if the people from Bhopal unite and fight together in a single forum, even though they may have separate banners, then they will be able to have a formidable strength. When Sathyu *bhai* and Appa (Rashida Bee) saw that I was operating alone, they called me. I joined them because one cannot function long alone and do as much work as we can united.

These young girls, by God, they will keep the name of gas tragedy alive

In our organisation, *Bhopal Gas Peedit Mahila Purush Sangharsh Morcha* we have formed a women's wing and men's wing. We have an executive president and a separate president of the women's wing: Zulekha is the president of the women wing. Executive president heads both the men and women. The men's wing is general, so we have made a general executive president and a separate president for the women's wing because these are some matters concerning women that however close a relation, whether you may even term it Father-daughter or brother relation, yet they will not be able to confide, so for the comfort of women, it was necessary to keep a woman (president).

Amongst the new generation, we can see girls coming up. If they do not marry or their husbands don't take them far away, then by God, they will keep the name of gas tragedy alive. We are seeing amongst these girls that in spite of being very young they have so much information and knowledge that we have to admit, they will lead well and take the people towards welfare. There is Yasmin, Sarita and 3-4 more girls whose names I do not know, who have both brains and spirit and I can tell you, they can capably answer the journalists, the media and if ministers ask, they are capable enough to answer. Water victims and gas victims are united, both are together. There were times that many wanted the water victims to take out their rally and raise their voice separately, but then we persuaded them that we all inhaled the gas together, at the same time and you have been drinking this poison for the last 8 to 10 years and that these skin allergies, handicapped children being born, the problems of breathing and decreased vision, are all because of the same reason. The effects which were researched concluded that all this was happening due to water and especially for women when their blood and milk was tested.

See we are fighting for truth and justice. You people are also like us, like our young generation. We only desire and hope that you people do not repeat this in a different place. You can campaign from your side that a company such as Dow should be opposed, closed down so that tragedies do not happen and such factories are not allowed to be set up. And if you make even one person stand up to these companies, we will think we have achieved a big victory.

Rasheeda Bee

Bhopal Gas Peedit Mahila Stationery Karmchari Sangh

Bhopal Gas Affected Women Stationery Workers' Union

Rasheeda Bee leads *Bhopal Gas Peedit Mahila Stationery Karmchari Sangh*, the Stationery Workers' Union, along with Champa Devi Shukla. An uneducated though politically astute and powerful orator, she has come to international attention through her world tours with International Campaign for Justice in Bhopal. She and Champa Devi jointly won the Goldman award for environmental campaigning in 2004, from which they formed Chingari Trust which supports both disabled Bhopali children and women fighting corporate crime.

Whatever work is given we will do it, but we need work

The gas leaked on the night of the 2nd December 1984 and on 16th December Bhopal was evacuated. I left for Suhagpur with my family and returned from there after 6 months. When I got back I found out about the employment that the government of Bhopal were providing for the gas survivors, but to start with I hadn't a clue where it was. I came to hear about Bharat talkies and the stationery and sewing centres so I went to register my name. They asked me if I would do the work of making and folding papers. People needed work to earn money to provide for health care, so I would do anything. So my name was registered and sent to the Collector's office for verification.

They registered 50 Hindu and 50 Muslim women. We would get to the work sheds at 10 in the morning and come back after 5 in the evening. We got Rs. 5/- per day whilst we were being trained. After a month we got Rs. 150/-. By the 3rd month they gave us some proper work and at the end of that month we were paid and asked to go home and just do what we had done in training, work from home and sell the finished products.

The women hardly knew each other but we were aware that we had not been trained enough to make our own products and sell them. After talking it through we decided that we should raise the issue, so the women asked me and Didi (Champa Devi) to speak to the Collector. Well I had never spoken to any man or boy since my childhood since this was the custom that I was taught and had always followed in the early days. The Collector, Mr Pravesh Sharma came to the shed on the day it was due to close. We spoke to him and gave him the women's views about how they had still not learned the skill properly and that they didn't want the shed to be closed. The Collector replied that training has been given as planned: the most they could do now was to help them to get a loan to set up a business and then they are on their own to make and sell the products. But the women insisted, saying 'how are we to manage all this when we haven't a clue of any of this process? We are not educated enough to carry out the work in an organised manner.' The Collector said we would have to go and speak to the Chief Minister and if the CM agreed to their demands then the shed would remain open. At the time, I wasn't aware of who the Chief Minister was or even what a Chief Minister did. That same day we walked straight from the sheds up to the CM's house and reached there at 3 in the afternoon. The guard at the gate told us that there is a proper time to meet the CM and that we should come back in the morning.

The next day morning at 7:30 in the morning we reached the CM's residence. Around 8 o'clock Motilal Vohra stepped out and met everyone and the person behind him collected the sheets of paper with requests written on them that people were handing over. But we had no sheets of paper. So we stepped out of the crowd and told him that we were gas survivors and that we wanted our work sheds to be kept open and that we should be given work. At least that way I would be able to take care of my father who was sick with cancer. We were asked if we were willing to work at piece rate and we replied "What ever work is given and available we are ready to do it, but we need work".

The next day the Collector received a letter and the women from the workshed were handed over to *Rajya Udyog Nigam* (State Industries Department). We were to get work from the *Udyog* and get paid by them at piece rate. At the time we didn't know what this meant. We were only interested in getting work and not who gave it to us. We sat for the whole month with very little work coming in. In April 1986 we got paid Rs. 6/- for the whole month of March. We were furious and questioned the authorities why we were not given our standard Rs. 150/- per month at which they told us that we had agreed on getting paid piece rate and that this pay was for the work that we had done. It wasn't until then that we discovered what piece rate meant. But we protested. We said "we did no work because we were not given any work, so we are not at fault". But we did go into work every day from 10 to 6.

They refused to listen to us and we were told that this is the amount that was going to be paid, we can take it or leave it. We protested. In our opinion, be it piece rate or not we should be paid the usual Rs. 150/- a month. We decided to refuse to take the money that was being paid. Mr. Gupta who was the accountant then, encouraged us in our decision when he said that if we did not take our pay but still keep coming to work the *Udyog* would be forced to increase the amount of work for us to do.

The Union created a sense of fear amongst the officials

The Chairman of *Udyog Nigam*, Manak Aggarwal also came and asked us to take our pay. But we didn't take our pay for three months. Every day about 15-20 women would go to *Rajya Udyog Nigam* and shout slogans and demonstrate, demanding work. After three months it was decided that we would be given at least Rs. 7-8 worth of work every day. We were then taught how to stitch registers, file covers etc and on the basis of this we were given work. The women sometimes managed to get Rs.12 for a day too. So pay would vary from some women getting Rs.75 to some getting Rs.200/- per month.

But we didn't stop demanding more work. Soon we were being given enough work to make Rs.300 – Rs.350/- per month. We would work like crazy with very little rest because each one of us needed the money to feed our children or for medicines.

Two years passed by and we continued to work in the same manner. Then in April 1988 we found out that over this period *Rajya Udyog* had made a profit of Rs. 4 *lakh*. By this time we had registered the Stationery Workers' Union.

It was around the time of the wages boycott that Jaan Nisar of the Communist Party suggested that we register a union under our names. He proposed forming a 12 member team where one became the chair-person, another a secretary and a third treasurer, the others being the working members. We were told that forming the union would help us fight with more strength.

We collected Rs 5/- from each of the women to finance me and Didi to travel to Indore and get the registration done. At the registration office we were given forms to fill in to register and to write down the names of all the members. We filed two forms one after the other and both were rejected saying they were not right. We returned to Bhopal empty handed. Then Mr. Yadav told us to meet a particular person at the registration office and we would surely get our registration done. So we went back to Indore, reaching there at 9 am and waited for the office to open at 10:30. Someone called Sharmaji then approached us saying that he had seen us sitting there for the last two weeks, what might be the problem? After discussing it with him, he asked us for our papers but said that it would cost Rs 200/- for getting the work done.

Seventeen days later we got our registration letter, dated 17 March 1987. What is more he had waived his fee of Rs 200/- saying that we were all women and his Rs. 200 could be compensated by someone else. The women tended not to tell anyone about anything, so no-one knew about what was being done to register the union. So when we showed the registration letter to *Rajya Udyog* and demonstrated that we were a properly constituted union, they were shocked and had to acknowledge our efforts.

The registration of the union worked well for us. All the letters that we put in were received and most of our demands were met easily. Registering the union had created a sense of fear amongst the officials. It was after this change that we started receiving work worth Rs. 300 – Rs.350. We would go and sit in front of the *Rajya Udyog* Chairperson's residence, block his car, and create havoc demanding work and more work. It got to the point that trucks would come with raw material at night so that the women had work to do when they came in in the morning.

The MD tried to buy us off rather than fulfil the law

Mr. Gupta the accountant for *Rajya Udyog Nigam* was very helpful to us and so was R.K. Yadav, who had been Chairman of the union at the Union Carbide factory and was later given a job at the District Industries Centre. Both these men gave us information or some advice every now and then. On the basis of this advice, we decided to write a letter saying that according to the laws of 1948 if twenty workers work together under one roof they come under the Factories Act and so they should be registered as a factory. Thus, up until now, the sheds had been run illegally and now it should be run according to the law and the 4 *lakh* profit should be distributed amongst the workers.

When B.K Tewari (Managing Director) received the letter he spoke to Didi and me and said "This is a very big demand, nobody can fulfil this demand. What can be done is that we will increase your rate of pay. For the work you are currently getting paid 10 *paise* we will give you 12 and for stitching the register for which we give you 25 *paise* we will give you 30 *paise*." He tried to buy us off rather than fulfil the law. But we were adamant. We told him that he can reduce the price that is being given for the work instead of increasing it but that we should be legally registered and given a share of the profit. We insisted that the work sheds should operate according to the proper law that has been made to run the state effectively.

He conceded that our demands were right but that he was not the right person to deliver it – although he also said that he should not be quoted on this! He told us that the only person who could help us was the Chief Minister.

When the women were told about this it was decided that afternoon that the next day we would gather at *Vallav Bhavan* (Secretariat) and protest for our demands. So the next day we went and sat down in front of the *Vallav Bhavan*. Satyam sahib was the Gas Secretary then and said that this demand cannot be met and that we should continue to work the way we were working or else we would lose this job too. We didn't take any notice and decided to stay put at the Bhavan. We hadn't planned to stay there nor had the women told people at home that they would be staying. A young boy called Raies told me "*baaji*, since you have come all the way here then why don't you sit right here and protest". It was not something that was planned. So this is how the idea came about for our first *dharna* at the *Vallav Bhavan*. No leader had ever sat there and protested before. Raies offered to get a tent set up, so in front of the *Vallav Bhavan* there we sat under a tree on our first *dharna*.

A little after five in the evening various family members started worrying about where we were. The asked around and found out where we were and came to the Bhavan and found us sitting there in protest. During those days no man or family member ever had a problem as long as the decision was taken by me or Didi and we were there with all the other women.

We just ate watermelon that night and the next day some women went home and each cooked food for 3-4 women and came back. Days passed by like this and still no response from the Gas Relief Ministry. After about 6-7 days of *dharna* we decided to go on hunger strike which we did in shifts. Every 24 hours a different group of women would go on hunger strike, 10 one day and another 10 the next day, and so on. Twenty days passed with us continuing the hunger strike like this. On our 22nd day we found out about the elections at Kharasia. Arjun Singh was the CM and he was standing for re-election, and it was very important for him to win back this seat. His position as CM was at stake with this election, and he was in Bombay at the time.

I started to realise that this is about saving the world

There were many people who were falling sick beside the Union Carbide walls and all around it. Why were they falling sick? Most of the women who I knew were from these areas where people were facing new problems. I met up with Sathyu and he told me about the contamination of the water. And after the reports in 1999 it was found for a fact that the water was indeed toxic. In one of the hand pumps black water started gushing out and every one went to see.

Greenpeace came in 2000 and it was after this that we in the *Stationery Sangh* joined hands with Sathyu. After hearing about the contaminated water, and from what I had learned over the years, I started to realise that this is about saving the world. What happened in Bhopal has already happened, but we need to join forces to stop it from happening again anywhere else in the world. I also came to know about the law that says the polluter must pay, which strengthened us all the more because we now knew that we had the law on our side. We found out about lots of things that were happening throughout the world from working with Sathyu. Then in 2004 the International Campaign for Justice in Bhopal was formed.

There had been a conference in Japan in 1996 which some people from Bhopal had gone to. There had been a mercury leak from a factory in Japan and many people were affected and children were being born deformed. People had been fighting for justice for 40 years. Earlier they got a compensation of $13,000 and then they got 13 *lakh* and now they have managed to get as much as 40 *lakh* as compensation. We got motivated further by this, if the people of Japan could fight for 40 years and get their rights then why can't we? So in 2001 when there was another invitation to Japan, I went with Pranay. We saw the situation there and the state of people. It was due to high amounts of mercury, which is also the case here in Bhopal. This meant that we could face the same problems that they have in Japan if we didn't do something about it and stand up against injustice.

Then in 2002 I went to South Africa for a conference regarding environment and its safety – the World Summit on Sustainable Development. There too I saw and heard more about Dow's atrocities. But many of these chemical companies were also present at the conference and this made me uncomfortable. What was the purpose of this façade at the United Nations in front of the prime ministers of all the world's countries? It made me doubt whether the UN was in reality ready to work, or wanting to work for the environment, when I saw that companies like Dow were a part of such a conference.

When I had the chance, I stood on the stage and put a question to Kofi Annan: why were such chemical companies at the conference and why can't the UN prevent them? These are the people who are the causes of some of the worst disasters, those who had ruined the land and forests of Africa they were all there. How can this be a conference for environmental safety if they were a part of it too?

Environmental issues just came up. How I found my way through them all, God only knows. Mostly I just thought it was right, that I could do something about it and did it. When the black water came out of the hand pumps, some refused even to touch it and there were other people who had to drink it every day. The people who came to test the water wore gloves. All this just pushed us to believe that change is necessary and the justice we were struggling for could save thousands of lives.

Anil Sadgopal had once said that it was declared on the 3rd of December 1984 itself that mothers who were carrying on that day, even babies older than a month in the womb, they would most likely grow up to give birth to deformed children.

This fight is not for the gas survivors or compensation but a fight for the world. And the fight is against the companies which even with all the knowledge are still spreading toxics across the globe. And to save the world from this the struggle in Bhopal has to spread across the world. Bhopal should be taken as a lesson to learn from, an example which the environment itself has given to people that we should not push our luck and try to change the simple ways of life.

The air is polluted in most parts of the world. It doesn't matter if they are a developed country or not. They have been equally affected by the growing industrialisation and the spread of corporations. Most of the products today have some chemicals in them and hence the fight today in Bhopal is not limited to the rich or poor but is the fight of the world for the environment. Because if the environment is not clean then humanity itself cannot survive.

Satinath Sarangi (aka Sathyu)

Bhopal Group for Information and Action
International Campaign for Justice in Bhopal
Sambhavna Trust

Satinath Sarangi (known as 'Sathyu') is from an educated, English speaking, left wing background. He was amongst the earliest outsiders to arrive in Bhopal after the disaster, one of the founders of *Zehreeli Gas Kand Sangharsh Morcha* which he later left and formed Bhopal Group for Information and Action. He operates on a global stage and has done much to internationalise the campaign, being a founder of the International Campaign for Justice in Bhopal. He is managing trustee of Sambhavna Trust which provides allopathic and ayurvedic medical care to gas victims and conducts research into the ongoing health impacts of the disaster.

The *Zehreeli Gas Kand Sangharsh Morcha* was founded within a week of the disaster in a meeting of social and political activists and social workers from within and out of Bhopal who had gathered to provide support to the survivors. Most of us were from privileged backgrounds and except one or two there were no 'victims' in that meeting. While the organisation and its three leaders were named, objectives and activities defined and plans and strategies chalked out with near total exclusion of the actual survivors, the *Morcha* presented itself to the survivors as a democratic organisation that encouraged and supported participation of ordinary survivors. Survivor activists participated in the *Morcha* as representatives of their individual communities and their participation was mostly sought in mobilising survivors for demonstrations and rarely if at all in important decision making in the organisation. As someone working in the community (as opposed to the leaders who seldom visited the affected areas) I was always troubled by the lack of democracy in the organisation but those were such firefighting times that there was very little time ever for discussions on organisational questions. The repression by the state and the practical need for secrecy in the organisation further legitimised the top down structure. Thus a few of us middle rankers were helpless when a group of out of town activists, all very dedicated workers, were falsely charged with planting bombs within the Union Carbide factory by the leaders and thrown out of the organisation.

As the person in charge of the health clinic run by *Morcha* I tried to create a space within the organisation where democracy was better practiced. Within the clinic, all decisions were taken collectively and based on consensus rather than on majority rule as was the practice in *Morcha*. When I questioned the basis on which one of the *Morcha* leaders took a large number of survivor activists to a conference of his political party, I was summoned to a special meeting and was thrown out of the clinic and the organisation by a handful of activists loyal to the leader. Not a single survivor had been invited to that meeting.

When a few of us later formed the Bhopal Group for Information and Action we decided we would not be part of any survivors' organisation but would support all organisations from outside. Our role though limited to gathering and sharing information and advising on strategic matters became critical for the several survivor led organisations that sprung up following the demise of *Morcha* as a mass based organisation. In our relationships with these organisations, which were even less democratic than *Morcha*, we did our best to empower the rank and file members and increase their participation in decision making within their organisations. The all powerful leaders of these organisations were never comfortable with our approach but they tolerated us because they needed us.

Predictably, these relationships broke down and resumed in different forms. After staying in the background and supporting survivors' organisations for several years, the BGIA decided to be more visible and work in coalition with survivors' organisations. While little changed within the organisations, we were able to establish democratic participation of the leaders of the organisations in the coalition and install a culture of collective decision making.

For the last six years now BGIA has been part of a coalition with two survivors' organisations both led by survivors themselves and much of the sustainability and strength of this coalition comes from the democratic practices that have evolved and over the years become part of our culture.

Issues of health and health care have remained among the most bitterly contested political matters of the ongoing disasters in Bhopal. For the survivors and those exposed to contaminated ground water, access to medical facts on deaths and actual and potential injuries including in the next generation are critical for their survival. They have indeed fought a very sophisticated and protracted political battle in the face of suppression, manipulation and destruction of medical facts and figures by Union Carbide and the state and central governments in India.

My personal exposure to the hard politics inherent in these 'humanitarian matters' first happened in June 1985 at the People's Health Centre, a joint effort of four organisations, including the *Morcha* of which I was a founding member. We administered sodium thiosulphate, by then a known detoxificant, to survivors and kept detailed records of the resulting improvement in each individual. After just 20 days of this work, on a midnight in late June, I along with more than twenty others including volunteer doctors were arrested and sent to jail on trumped up charges of attempt to murder. The information we were generating regarding the administration of the drug could potentially have demonstrated that the personal injuries caused by Union Carbide were far graver than was admitted, and lay the basis for greater liability and claims. This was therefore politically unacceptable for the Indian government. Two years later a copy of a report on exposure-linked damages by a government agency was found in the little office of Bhopal Group for Information and Action during a police raid, and we were charged with spying.

With respect to issues of health care the unprecedented (both in magnitude and complexity) ongoing medical disaster has laid bare the limitations and failures of the synthetic drug centred hospital based approach in providing appropriate health care in Bhopal or for that matter for chemical victims anywhere in the world. From 1985 to 1995 while I worked through BGIA on publications of newspapers and newsletters, generating information, taking legal actions and supporting survivors' organisations I was never far from matters of health care. Almost every day I would meet a family or two where one or more persons were suffering chronic exposure induced illness and had got only temporary relief, if at all, despite prolonged treatment at the government hospitals. The studies we carried out showed that treatment in the hospitals was potentially causing more harm than good to the survivors and that the ongoing disaster had become a windfall for multinational pharmaceutical corporations who had a captive market in Bhopal.

The Sambhavna Trust and Clinic that I helped set up in 1995 came out of the grief and frustration I endured for 10 long years as the health and health care situation of the survivors steadily deteriorated. We saw the failure of the state government as an opportunity to create and legitimise an alternative approach to health care that integrated non toxic and drug free therapies and laid stress on health education and community participation. While we are aware that with our limited resources we can only provide health care to a fraction of those in need of medical attention we have been able to influence the dominant system of health care in significant ways.

We have always taken the approach that health care cannot be provided by medical specialists alone. In many ways the survivors are the experts in their conditions. Not only is it vital to use therapies which can be controlled by the recipient, it is also important to gather the people's knowledge for the purposes of research. At Sambhavna we have been gathering information which has been told to us by survivors and have pioneered the use of 'verbal autopsies', so that the bereaved are given the dignity of an assessment of cause of death when their loved ones die without reliable medical contact.

Sambhavna has been made possible because of financial support from a few organisations and very many individual donors throughout the world. Whilst collecting international funds has been seen by some as controversial, the issue has always been an international one.

The disaster was caused by a multinational corporation and was headline news in newspapers of nations all over the world. Worldwide there were spontaneous offerings of sympathy and support for the victims and anger against the perpetrators of the massacre. As part of *Morcha* I was aware of several offers of solidarity from international organisations and individual activists. In several countries where Union Carbide had a large presence, environmental, human rights and anti-corporate organisations along with workers' and students' unions formed coalitions in support of the Bhopal survivors.

While the leaders in *Morcha* were aware that a corporation that operated worldwide could not be confronted without international solidarity they were wary of being vilified by the government for their foreign connections and failed to make use of the spontaneous outpouring of international support. The Bhopal Group for Information and Action built upon what remained of the international support two years after the disaster through its periodic newsletters and much correspondence. It became part of the International Coalition for Justice in Bhopal (ICJIB) that was formed mainly through the efforts of Ward Morehouse in the USA who set up the Bhopal Action Resource Centre immediately after December 1984. The Coalition had members from USA, UK, Ireland, the Netherlands, Japan and other countries. It grew in strength and effectiveness with the campaign tour of four countries (USA, UK, Ireland and the Netherlands) by three survivors and a BGIA member in 1989 and also through the organisation of the Permanent Peoples Tribunals in Bhopal, Yale, Hong Kong and London from 1992 to 1994. The International Medical Commission on Bhopal had 15 doctors from

11 countries visit Bhopal in 1994 to make an assessment of the medical situation of the survivors.

Later, Bhopal survivors' organisations and the BGIA collaborated with Greenpeace International on several projects. However, this association was ever fraught with tensions because Greenpeace's corporate structure offered no space to the needs and opinions of local organisations. The International Campaign for Justice in Bhopal (ICJB) formed in 2003 was a significant improvement on the ICJIB in terms of contacts and solidarity actions and has a much wider base particularly in USA. Whilst all members throughout the world contribute to information sharing and campaign decision making, the campaign distinguishes itself in having the Bhopal based survivors' organisations as the final arbiters of all decisions influencing their lives and struggle.

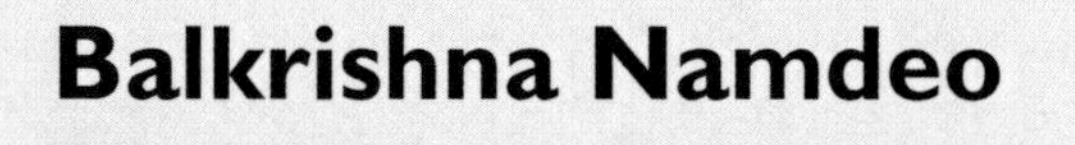

Balkrishna Namdeo

Gas Peedit Nirashrit Pension Bhogi Sangharsh Morcha
Gas Affected Destitute Pensioners' Front

Balkrishna Namdeo moved to Bhopal as a child to go to school and soon became politically active on behalf of the most vulnerable and oppressed. He took up the cause of the elderly and disabled who were entirely dependent on state pensions, and has been leader of the pensioners' campaign *Nirashrit Pension Bhogi Sangharsh Morcha* since before the gas leak.

Social security pension scheme was initiated in the state of Madhya Pradesh in the year 1981. Our organisation was formed to deal with the difficulties that the beneficiaries of this scheme (old, disabled and widows) were facing. In the beginning we noted that there was only one place in all of Bhopal that dealt with all the pension related work and people faced a lot of problems with that. So we decided to organise people to fight against government apathy and that was the beginning of the organisation.

When the organisation was founded there were a lot of people involved. Many of them have since died, some have moved on or shifted to other places and some who are still with us. After the gas leaked in 1984 and the whole city was in the grasp of the aftermath we decided to include the cause of the gas victims into our organisational agenda. Now we work on both issues.

We chose the pensions issue because there was nobody or no organisation to fight for them. There were labour unions, trades unions, women's organisations, youth organisations, employees' organisations etc. but the old had no organisation to fight for their cause. The group ended up with a very long name – *Gas Peedit Nirashrit Pension Bhogi Sangharsh Morcha* – and we have tried to shorten if for some actions. But we always say that the most important part of the name is *nirashrit pension bhogi* (destitute pension entitled) because we are fighting for people with no other source of income. A decent pension would lift our members out of destitution.

The *Nirashrit Pension Morcha* has always been a peoples' movement although people join and leave all the time. There are very few people who have remained with the organisation over the years. Most people come with problems related to pension, BPL, water, hospitals. They come and get these resolved and then become part of the movement. A lot of people also leave because of old age, health problems and other practical issues. This is an ongoing movement and we have thousands of members even today, gas victims and pensioners. There are members outside Bhopal who are not gas victims and many gas affected pensioners who are completely dependent on their pension. There has been no special pension for gas victims.

I was born in Khurai, a small town where it was impossible to get higher education. I was also orphaned at a very young age so I had no family support. When I was around 15 years old, I moved to Bhopal to further my education. But I also had to earn a living and pay for my college fees. In 1978 I started a small groceries business in New Market. I went to college during the day and ran the shop in the evening.

Later that year, the City Corporation banned our business in New Market. They would remove all the temporary shops on the road and also seized our goods. This was a great challenge for me because it was a matter of my livelihood and education. So I decided to organise people like me in the market. We formed a union of all the hawkers called *Futkar Subzi Vikreta Sangh* (street vegetable sellers' organisation). I always believed that one person can guide a struggle but he can never win it alone, an organisation is necessary for that.

We were a group of 17: 12 women and 5 men who formed the organisation. RSS was the ruling party in the government at that time under the leadership of Virendra Kumar Saklecha. The union went on a hunger strike for a week against the city corporation. It was our first demonstration and we were frustrated that nobody was listening to us so we decided to do something powerful and we *gheraoed* (surrounded) the head of administration I.S. Rao. One of the women in our group fainted and the men caught the adminstrator and beat him up. We were all arrested after this and it seemed like everything had come to an end for us. The Corporation banned everyone from trading in the area.

I had no understanding of politics at the time. The protest had been supported by Congress workers but they abandoned us when the trouble started. When we were released from prison we approached various political parties but nobody helped us except the Communist Party leaders. I admired their honesty and integrity, they were the only ones to help us and expect nothing in return. That's when I realised that the only people who can really help the poor are the communists and this was when I got interested in communism because it appealed to my way of thinking. So I joined the Communist Party of India (CPI). CPI members helped me form the group.

The fortunes of communism internationally had an impact on India and also on Bhopal. The communist movement had been very active in Bhopal, and there was one CPI MLA from Bhopal until 1978. Newer party workers adopted different ways of working. A lot of effort went into electoral politics which had no impact. Parliamentary elections led to alliances being formed which demanded big compromises with Marxist ideology, such as with Chandrebabu Naidu in AP. The vigour with which we fought back became gradually diluted. I thought we were going to see a national revolution, a big force that would fight for justice but I realised that this was not going to happen through the Communist Party. So eventually I quit the CPI.

I used to read a lot. I read about other leaders like Ambedkar who inspired me to do a lot of things in life. Ambedkar spoke about the importance of education which made me prioritise education in my life but I also had to use this education for the improvement of my society. I had to share it and put it to good use.

The Pensioners' struggle

I was always a social activist with ordinary working class people. I had people coming to me with their problems, usually pension or some other official problems, and I used to take them to the collector's office and help them out. But I also knew the importance of having an organisation that would fight for their cause. Once the organisation was formed the same people who were once dismissive about the problems of these people were suddenly very responsive. Our biggest victory has been to get the government to authorise banks for pension disbursal instead of a single office in each district. We raised the issue in Bhopal and the rule was implemented throughout Madhya Pradesh.

When the pension started getting disbursed through the banks, the dates of pension distribution kept changing every month according to the convenience of the banks. The recipient has to make several trips to the banks just to get the date. Most of the beneficiaries are illiterate and they confuse or miss the dates, and if they miss the date for more than three months they stop getting their pension. So it was essential that they were aware about the dates and it was also necessary to work with the banks to encourage them to fix a single date. Our organisation fought first to get the banks to distribute pension and then to get a fixed date. The Supreme Court also ordered that the pension should be distributed before the 7th of every month.

The state government scheme is to provide social security pension to widows, old and senior citizens and the disabled, all who are unable to work and who do not have family support. Among the disabled, a pension is given to school children between the ages of 6 and 14. Children above the age of 14 who do not attend school have to be from a family with a BPL card in order to get the pension. The pension forms were very difficult to understand and complicated to fulfil the requirements. We demanded simplification and this campaign was successful. We have also contributed to the changes to pension entitlements which happened in 2003, to include widows under 50 and women who have been abandoned by their husbands without formal divorce.

The amount of pension was Rs.60 back then and we fought for an increase to Rs.200. This happened gradually, from Rs.60 it increased to Rs.100 then Rs.125 and then finally to Rs.150. The Government of India gives Rs.200 per month to old people over the age of 65 in which an additional Rs.75 is contributed by the state government. So the Social Security Pension is sponsored by the state government and the Indira Ghandi Old Age Pension is sponsored by the central government. However, although pension has increased, it has not increased in relation to the market price of basic commodities, and the state government hasn't increased the Social Security Pension since 1996. So our struggle for a better pension continues.

Now our main demands are: to link the pension to the cost of living; for the same access to government rations for pensioners as for BPL card holders; for free medical care for all widows and old age pensioners; and to establish a National Commission on Old People's Welfare.

The *Zehreeli Gas Kand Sangharsh Morcha* was the only organisation in existence immediately after the gas leak which was formed by educated people from outside Bhopal as well as by locals. The outsiders were all trained, they had been part of movements before the disaster but we were all new to the whole concept, we had only been part of smaller fights. We slowly learned how things worked in a movement, things like whom to target and hold responsible for what. These people from outside were well informed and they knew how things worked in a situation like this. When we spent time with them we learnt a lot about the way things worked in the system. The *Morcha* broke up and the members formed different organisations, but a lot of people left because they probably felt that the locals were in a position to handle things on their own. Today the movement is being managed by the local people, we may receive outside help but the struggle has to be kept alive by us.

There was also a CPI led organisation called *Gas Rahat Aur Punarvaas Samiti*. Because I was with CPI I was part of this organisation as it was decided by the party that the members should work with this organisation. However, I felt that the organisation could not initiate a grass root struggle. One of the problems of the struggle in Bhopal is that although there are many organisations, most of them do not work with the grass roots. Some receive government aid to provide training so their focus is to show immediate results out of the money they get. Such organisations can only benefit a small group of people and not the masses. Employment can never be provided through NGOs, there has to be a change in economic policy.

India should have its own economic agenda and policy. We have the resources to do that but we are just not doing it. Why is there so much illiteracy in India, why have we failed to provide good medical care, why is there still unemployment? There are so many issues and companies will not come and change all of that. Government policy is what can change it.

When multinational companies come to India, Indian laws are not enforced. Indian law could not even get close to prosecuting Union Carbide and Anderson let alone punishing them. So if companies get into India on the pretext of employment then they will only cause destruction.

If these companies want to invest in India they can do so without challenging our sovereignty and within our laws. They should not be allowed to invest, earn profits and leave a mess like Bhopal behind. When Enron came into Maharastra what kind of employment did it provide, what kind of development did it bring to the region? Nothing!

Women have always been at the forefront of the struggle. It's been part of the history of Bhopal, a *Begum* (queen) ruled this city when it was rare to see women rulers. Women played a vital role in the freedom struggle and in the struggle that ensued after independence when the states were being merged. Comrade Mohini Devi Srivastav was a prominent leader of these times, she shaped both these movements. There were other women with her like Basanti Devi, Shanti Devi, Akthar Jahan Begum etc. So there is a history to this place and its women, how could women not have been in the fore front of the movement?

Women take care of house work and they also take responsibility of work outside. Men do not join the movement in the same way as women. I think about this a lot but I don't know the real reason behind it. It defeats me. You will find a woman fighting her way through a queue at the ration shop with a child in her hand. You will find a woman fighting the system anywhere. Women are at the forefront of all movements. So I believe that the women are more aware than the men.

The men do join but they always give excuses for not participating like "I do not have the time". There is no long term participation, the kind of commitment seen among women. Most of the women who work with me are old, they have a lot of family problems, but they never display any weakness when it comes to fighting. When I formed the organisation people were sceptical. They said it would not last because even big trades unions and organisations break up, and this was after all led by old women. It's been more than 25 years now.

I believe that only the most marginalised and exploited have the fire in them to fight. These women face a lot of difficulties and they need to be organised. They need me and I need them. I identified the most marginalised members of the society, the old, disabled and widows and organised them. So the old people not only lead the struggle but they also show the way.

It's not just about leadership. A movement requires a lot of planning and that cannot be done without education. It is the first condition to run a movement but it does not mean that an uneducated person cannot be part of a struggle or that an illiterate person is less aware. A lot of insights of the uneducated prove to be much better than the educated. I am able to offer leadership to the women in this organisation through my education, but they lead me in a lot of other struggles. Ram Pyari Bai, one of our prominent members is 85 years old but I feel weak in front of her sometimes. I am merely helping them, they are leading and showing me the way.

Education is important. It is important for there to be educated people in a people's movement. But it is our misfortune that even after independence there is a high percentage of illiterate people in our country. India has the largest illiterate population in the world. So the government policies are to blame for all the social problems in the country.

The rich cannot fight as efficiently as the poor, as is the case in many movements. Only the poor and the working class fight, only they can initiate a movement. There is a section of victims in Bhopal that is rich and they have received as much compensation as the poor. In fact they got much better compensation than the poor, 95% of the gas victims received nothing more than Rs.25,000. I know of an additional collector and his wife who received Rs.50,000 as compensation. They live in a posh part of the city in an independent bungalow.

So people who had access to resources and were aware about the system manipulated it and got better categories for themselves. In order to obtain compensation, it was necessary to have a medical check up and then be allocated to a category: A, B or C on the basis of the doctor's assessment of the disabling impact of the gas. Richer people were able to manipulate this and be allocated to the higher categories, whereas 95% of the poor were allocated to A and B, the less severe categories. The whole compensation system was wrong and the worse aspect was the medical certificates which were used as a basis for compensation.

Gas victims should be provided with community level health care. The big hospitals should only be visited when the community health centre can't treat a person. This should be set up under the government scheme. NGOs like Sambhavna Trust cannot provide care to all the 36 wards. What I propose is a comprehensive system that looks into all the problems of the entire affected population. The NGOs can initiate this but it can only be done with support from the Government. Government hospitals could also initiate this. All government hospitals should provide free treatment to gas victims, not just the special gas hospitals. A National Commission would be able to look into all issues affecting the welfare of older people.

All governments judge the strength of organisations and then they also try to assess political gains from association with them. So NGOs exist as long as they get support from the government. A committee of grass root organisations should also be formed and their opinion should also be sought on various issues. There is no effort from the government front; we have to work so hard for a meeting with the Government. The present state government has formed committees but they consist of their own party members and they have refused to hold any talks with NGOs. Governments have earlier interacted with organisations and also sought their opinion on various issues but that process of public participation has been ended. No governments take the issues of the gas victims seriously any more.

The Government does not feel threatened by the gas organisations. Unless a united front is formed it is very difficult to achieve that respect from the Government.

The fight for justice in Bhopal will go on as long as justice is not done. There is no room for defeat because this is not a political fight for supremacy. There may be delay but justice will be done. The movement belongs to all gas victims so there has to be participation from all sections and age groups. The next generation has to make an effort to learn about the gas disaster and families with young people have to educate their children about the disaster. I would say that each person has to participate in the fight for justice and stand up against any kind of injustice; it does not need to be for the gas victims alone.

Badar Alam

Gas Peedit Nirashrit Pension Bhogi
Sangharsh Morcha
Gas Affected Destitute Pensioners' Front

Badar Alam was a child when the gas leaked, then as a young man became active in the Pensioners' campaign, supporting Namdeo. Less active since starting a tailoring business he remains a committed supporter. A reflective and compassionate man and a skilled classical singer.

The struggle started from the day of the gas leak. The organisations formed later but the people of Bhopal initiated the fight for justice from the very first day after the leak. There was an action on that day but it was not very successful because people were still busy putting their life in order and they were also confused about the course of action. Union Carbide was something that not all Bhopalis knew about but overnight the company became notorious all over the world.

I remember that night vividly. As I talk about it I go back to that night. Not many people knew about the existence of the factory and the ones that knew about it had no knowledge about the poisons being manufactured inside the factory.

I had the motivation to fight from that day but I was very young to have started anything on my own. I was only 14 years old when the disaster happened. I used to attend meetings where I saw these leaders and was inspired to join. That's where I met Namdeo-ji and I got involved with the *Pension Morcha* when I was 16 or 17. I could see that Namdeo-ji was fighting for the rights of the gas victims and also for a better pension for all.

The *Pension Bhogi Morcha* is a very old organisation. It was formed before the gas disaster and Namdeo-ji was already working on pensions. Since the gas leak he gave the organisation a new direction and included the issues of gas victims into the agenda. The members were already fighters and were used to this kind of work, they did not need to learn anything new, they were already fighting and this issue showed up. The existing organisation *Nirashrit Pension Bhogi Sangharsh Morcha* (Destitute Pensioners' Front) took on the gas issue under the banner of *Gas Peedit Nirashrit Pension Bhogi Sangharsh Morcha*.

I was not working so had a lot of free time and I felt I could contribute something, so I joined the organisation and started working with the pensioners. I joined the cause of old widows because I felt very strongly for it. I liked the work. I would have done it for my ageing parents. A lot of people have joined Namdeo-ji. I have accompanied him to other parts of Madhya Pradesh and witnessed rallies at least 1,000 strong. His rallies in Bhopal also had huge participation. I also began participating in all the rallies, demonstrations, actions that the *Morcha* organised.

Decisions in *Pension Morcha* are made by all the main workers: Namdeo, Tara Bai, Ram Pyari Bai etc. The issues are then discussed at the weekly meetings, no separate meetings are held. Meetings happen twice a week, Thursdays are reserved for members from Madhya Pradesh and Sundays are for members from Bhopal both gas victims and general pension claimants. For an action, first an issue is chosen. Once chosen everyone's consent is sought at the meetings and members express support by a show of hands. If a majority of the membership present are in agreement then the plans are executed. A lot of work goes into mobilisation of the crowds especially in today's scenario.

There were a lot of other organisations that were working but I got in touch with Namdeo-ji initially and started working with him. There were not so many differences then between the groups. Sathyu *bhai* worked with all these local organisations. There were many organisations that emerged, like *Zehreeli Gas Kand Morcha* etc. There were a lot of action groups that had no name, groups of gas victims who came together. When a disaster like Bhopal takes place educated people with experience come in and form these various organisations, giving a new insight to the issues.

All these organisations worked together until 2002 when we went to Bombay for the Dow action, the issue of Dow accepting liability for Bhopal. I was in the group. We were all together: Sathyu *bhai*, Rasheeda Bee, Jabbar *bhai* etc. They spilt up after 2002 for some personal reasons which I am not aware of. There didn't seem to be a fight or misunderstanding.

They all have different working styles and none of them can compete with the other. All the local leaders like Jabbar, Namdeo and Rasheeda Bee had their own abilities and they were all doing their own work according to those abilities. People from outside like Sathyu *bhai*, Taranjeet, Deena a lot of other people came in later. The only person who has spent most of his time with the people here is Sathyu, he was also in touch with everybody and he also worked with all the leaders (Jabbar, Namdeo) but they never thought that he would make his own organisation or trust. Sathyu's role was to connect people and bringing the organisations together on a single platform.

The organisations got together only at the time of the anniversary or for a rally to Delhi that happened too rarely. The differences and quarrels were on various things including authority: each person wanting to demonstrate his/her authority and the loyalty of their followers. It was these small issues that led to splits.

Namdeo-ji's working style is very different. He has made a lot of sacrifices. He never started a family of his own because he has prioritised the cause that he believes in. He has not jumped into the field after the gas leak, he was already working for a cause before the leak and he included this issue into his agenda. He does not have any other mission in life but to help these old women who come to him.

Namdeo-ji and Jabbar *bhai* are educated but the kind of education Sathyu has had, his English speaking skills, have allowed him to interact with foreigners who are constantly with him. There was nobody like him in Bhopal. He was an engineer and he was constantly in the company of foreigners. He benefited from this and so did the gas victims.

Rasheeda Bee joined him and benefited because her education is not that great. It was because of Sathyu that she has progressed and become what she is today. Sathyu has had an important role in connecting all the groups, he has also played a vital role in introducing the Bhopal gas tragedy and its victims to the world. He has worked with all organisations, met almost every gas victim and been to all gas affected wards.

The leaders like Jabbar, Namdeo, Sathyu and Alok Pratap who formed these groups are still fighting while many of the people of Bhopal have gone to sleep. Many Bhopalis will line up when there is an announcement for compensation but they will not come out to fight for their rights. We have to thank these organisations that they have kept the fight alive and tried to keep people active in the movement.

It was not always like this, people were really angry for the first 10 years, crowds would come together on various issues. There were massive turnouts for meetings compared to today when only compensation draws a huge response. The involvement faded out because people did not see any positive outcome to their struggle. There has been a new wave of motivation after the second compensation disbursement because they have pinned their hopes on more such compensations in the future. It is difficult to remain motivated for 25 years without seeing results. Twenty five years is a long time. Those of us who were children at that time are adults now.

But I also feel that those of us in the organisations have made some mistakes that have led to the lack of motivation and awareness among the victims. There is a reason why masses come out only over the issue of compensation. We did not treat them the way we were supposed to at a given point in time. But also people are too dependent on organisations.

Women are in the forefront because of men's attitude. Men love to kill time and do things like smoke, gossip, drink tea and talk. Men possibly consider it shameful to be part of a struggle, it knocks their pride – and I say this as a man. Initially the participation was equal among men and women, the men had to drop out because they were all daily wage labourers and other workers and they had to make a living. The women were relatively flexible and they had the time to participate. Today participation has reduced among men and women.

Not many women gave up the *burkha*, in fact they fought with them on and a lot of the foreign media was attracted to this because they were impressed with the women who were fighting with their veils on which was a big deal in the Muslim community. There is no relationship between the veil and the movement.

I would also add that the state government and the central government have fooled the people of Bhopal for the past 25 years. The gas disaster was a boon for the political parties because it was an added source of income. The compensation money was diverted for other government expenses like purchasing government vehicles or construction of new government buildings. You should visit the ITI centre made for the economic rehabilitation of the gas victims. The huge halls that are supposed to be training gas victims lie empty, there is absolutely no work being done on that front, the whole system is merely a source of income for the employees hired there.

Dow should clean up the site, but since the case against Dow is always delayed then it could be done by the Government in a systematic way and after making sure that there is no risk of accident. The waste has been lying in the state government's custody for so many years. Things could have moved if the state and central governments had the political will. However, Dow is responsible just like a person would be who is harbouring a murderer.

The state and central governments are mostly responsible because they never took any concrete steps to resolve these issues. If the Government is proposing to set up factories it should make a plan and share the information with all the people of the area so that they can make an informed decision. These factories provide jobs and improve the standard of living but they should be allowed only after making sure that they are not making anything dangerous. It has to be done with the consent and support of the local people.

My motivation is the future generations; I want the central and the MP government to work out plans to provide these children with appropriate jobs, medical care and rehabilitation for their future. Compensation was a one time thing, there had to be a long term plan for the victims who were children at the time of the leak. The focus was more towards medical care and rehabilitation but education was ignored, a school for the gas victims should have been built. A secure future is more valuable than a one time lump sum compensation.

There has been absolutely no effort to make a school for the gas victims when the land was available. This would be an extremely positive move in the right direction to secure the future of the children; it should be dedicated to the gas victims which will also ensure that it remains in the memory of the students and because it is dedicated to the victims it will also have to be part of the curriculum.

Justice for me would not be Rs.1 *lakh* or any amount of compensation. Justice for me is punishment to the guilty. The government has to initiate action in this direction, and it has to prove its loyalty to its people. Another thing would be proper and long term medical care for victims, the existing gas relief hospitals are in a bad state. It would be foolish if the people give up fighting just after receiving compensation, the only victory for me would be punishment to the guilty that would be the victory of all the gas victims. It will send a strong message to the world.

There will be no defeat if the law is allowed to take its course; if other means are used to influence the verdict, then we could lose. I would not rule out foul play given the state of things today.

I have learnt that a person who cannot take a stand or fight for his rights is not human. The gas disaster has taught us to fight in the event of a disaster. If a similar incident were to happen elsewhere we have the confidence and experience to start an organisation in that place to fight for the rights of those people.

Hajra Bee

International Campaign for Justice in Bhopal

Hajra Bee was initially involved in a communist affiliated trade union at the rehabilitation work sheds, which is no longer in existence. More recently she has become active in International Campaign for Justice in Bhopal. She is a fearless campaigner, energetic mobiliser and respected member of the leadership team.

They came here, inflicted pain and suffering, killed us and left

I was born in Serwasa, Madhya Pradesh (200kms from Bhopal). My father was a farmer. My mother and father separated and I was brought to Bhopal by my mother at the age of 3 years. I was married in Bhopal in 1973 and my in-laws are from a town called Pathari in Vidisha District, Madhya Pradesh. I lived with my husband's family in Pathari until the birth of my first daughter when I was 16 when we moved back to Bhopal. I had three more boys before the gas leak. I divorced my husband in 1991.

I used to make *beedis,* my husband was a daily wage labourer (manual labourer), his union was in Jummerati (in Bhopal). He was part of the Communist Party and we led a very decent life. A lot of people had already begun settling in the area beside the factory and previously it was primarily occupied by the *banjaras* (nomads). These nomads were brought here to construct the Union Carbide factory and we never interacted with them. A few poor people discovered this unused land and slowly the houses began coming up. My sister-in-law and her husband made a house here. They asked me to make a house here. I could not have made a house at that time with small kids. So I bought a house for Rs.1,000: it was a gift but I had to pay a nominal amount. Later the government was allotting plots in J P Nagar and our house was officially registered in my husband's name.

When the gas leaked it poisoned the area close to Union Carbide: J P Nagar, Risaldar Colony, Rajgarh Colony, Shakti Nagar, Kainchi Chhola, Chandbard, Railway Station, all the places where the poor live. There is certainly a reason why only the poor were allotted land near a dangerous factory like UCC. The direction of the gas was on this side, the gas spread to all these places and people died here. They were all poor people: why? Maybe because the educated and well off people knew about Union Carbide and understood what MIC was so they never came here. Only the poor came so close to seek shelter and they built huts to live in and then the gas disaster took place. The upper classes never came here they went to far off places, the VIP areas, the educated areas: this was through cunning.

Fighting is everything

I started being active after the disaster but before that I was a house wife and had no knowledge about the outside world. I discovered the power of standing up and fighting for rights when I fought for compensation for my son. My second son Mansoor Ali was left behind in the house on the night of the disaster and he was severely exposed and is suffering from gas related ailments. He was diagnosed with Blood Tuberculosis due to MIC exposure. At the hearing to decide compensation I told the judge in the court room:

"I do not want the money, or my son! I want my 4 year old son like he was before the gas leak. I do not want a gas exposed kid. I do not want a sick son. This is the Government's mistake that in the 1970s the

Government of India permitted the foreign companies. They came here, inflicted pain and suffering, killed us and left".

The judge stared at me and asked "what do you expect?"

I said "a doctor's report".

One month later the doctors were called. The judge did not call the doctors who had been treating Mansoor Ali, he called Dr. Khare. Mansoor had a big file at that time, there were 11 x-rays in it because the hospitals were like his home. My son's childhood was lost, he could not read, write or play, was bed ridden all the time and a bottle was attached to him. How far can a mother take this? This was not the question of Hajra Bee's son but a lot of other mothers who went through the same thing.

The judge read Dr. Khare's medical report in which Mansoor Ali was declared normal. I got really mad at the judge. His own doctor had diagnosed him with Blood TB! If he was normal then why did he have to go through the 6 monthly courses, why does he keep being admitted to the hospital? If a child is normal then he has to be playing and frolicking at home so why does he remain in hospital? Then the judge closed the file and transferred it to the head office. I received Rs.55, 000 from there. I got some hope from this, that in fact fighting is everything, it's only through fighting that a person can claim her rights.

We stumbled upon the idea of forming a union

My first experience of activism began in 1986 or 1987 at the tailoring centre where I was working after the disaster. The centre was set up by the Government to employ women victims of the disaster. People at our centre were given the contract to stitch uniforms for government school kids but they were asked to do two stitches on the seams whereas the other centres just did one. This ate into our profits.

This is when we started speaking for our rights, I got the strength and a lot of women joined me and we demanded our rights and we won the single stitch fight. Then we stumbled upon an idea; there was a communist leader named Babulal Naagar whose relative was our colleague so she gave me the idea of forming a union. So Babulal Naagar helped us form a union and then we were registered and recognised. We were called *Gas Peedit Mahila Udyog Karamchari Ekta* (Gas Affected Women Workers United) union. About 450 women joined us in our work and our fight. Then we began taking up all these various issues, like if women are fired then we would protest and have them reinstated or if someone's products get rejected or if there are problems with the supply of raw materials.

Then suddenly in 1993 our sewing centres were shut down without permission. They said that the stocks of raw materials had run out and new stocks would arrive within 15 days and the work would start again then. We made many trips to the centre, we failed to receive the raw materials and those centres were shut down. Following that we *gehraoed* (surrounded) the *Rajya Udyog Nigam* (government department of industry) office which was of no use either. We had young children to look after and there was hunger, we could either fight or find food for two meals a day or take our sick little children to hospitals or should we look after ourselves. That's when I lost all hope.

From the time that Babulal Naagar helped us form the union, I was with the Communist Party and I participated in all their events with women from my *basti*. Pramod Pradhan was in charge of the Communist Party and when he got married to Sadhna Karnik who was active in the *Zereeli Gas Kand Sangharsh Morcha*, Pramod Pradhan started working with the *Morcha* and ignoring our union. We were being ignored because we had no one educated enough to guide us.

Slowly our group disintegrated and the money we had saved through the monthly receipts was taken away in parts from us on the pretext of miscellaneous rally expenses. All the union women who were communists asked for accounts but they were never given. We brought this up at the meetings but it was ignored by Pramod Pradhan, which really hurt us and I quit the Communist Party.

Everyone is consulted

When I was betrayed there and my views were not respected I quit. I would have been with the same Communist Party today. My ideas and views were ignored there, my complaint was rejected so I left. I lost my money and nobody listened to me, I quit the Communist Party. I do not want to be in a place where I am not heard.

I joined ICJB – I knew Sathyu *bhai* for a long time from the time he was running Sambavna. There was a consultation meeting near the statue when Appa (Rasheeda Bee) and Didi (Champa Devi Shukla) were going to America and they announced that there will be a *dharna* here and they were looking for participants. So Sathyu and Rasheeda Bee called me and my neighbour in J P Nagar to be part of the *dharna*. They had raised the Dow issue at that time after the water was declared toxic after tests. At that time the clean up demonstration at the factory site by Greenpeace also took place. That's when I joined them and my interest grew from there on.

There is a difference in the way things are done in ICJB and the way they were done in the Communist Party. In ICJB all the issues are brought out and everyone is consulted before making any decision.

There are communication issues – for instance some times I cannot grasp certain issues so I just wait and get a clarification later. When communication happens in English I feel a bit left out and I am also not sure how much of the matter actually gets translated into Hindi. I have that feeling of being left out of the issues.

When we have to gather support for actions, the *Padyatra* (long march which took place in February 2008) for instance, then we go into the *bastis* and give them the facts of the case and tell them the truth, we do not believe in misguiding the people. We have to make people understand the issues in different ways in order to ensure their participation. Irfan *bhai*, Nawab *bhai*, Mira and I are mobilisers. We visit the *bastis* every day, but on rare occasions leaders like Rasheeda Bee, Sathyu, Rachna and Champa Devi also go to *bastis*.

Everyone called out for Hajra

Even in 1992 I campaigned a lot, this was at the time of the '92 riots on the Babri Masjid and Ram Temple issue. During the riots every person called out for Hajra whether it was to stop police vehicles or to get military assistance or if there is starvation then to find out what the Government is doling out and is not available here. The rich areas received packets of bread and milk, free onions and potatoes and when the same trucks came to us they started to charge money. This was during the Babulal Gaur regime. The Government sent trucks and they used a measuring box to measure flour, rice, onions and potatoes at the rate of Rs.5 per measure. When people had no work, there was a curfew, there were riots and the sufferers were the poor. So that was the time when Hajra stopped those trucks and lifted stones to break their windscreens to convey to them that nobody had the money to buy the food so they should leave. If you want to dole it out then it has to be free or we will starve to death, we cannot pay for it.

Then the police atrocities started. They would raid homes after midnight at 1am or 2am. They would knock on our doors and pick up the men and imprison them leaving the women and children behind. Then one day I got really angry, I saw three military jeeps passing during curfew time in the night and I shouted and gathered the whole Nomad Colony so that we could get military help. When we told them, the police began releasing the men and they reduced the number of raids.

I lost my ability to tolerate my husband's bad habits

My husband began cheating on me and drinking and staying away from home for weeks while I managed the family. This was unacceptable so I decided to disown him.

He would tell me that he was going to be out for a day but would not return for 15 days at a stretch and only I know how I managed to look after the children. He was a bit astray even before the break up but our life was going on at that time. His behaviour got worse after the gas leak and he also began consuming alcohol.

During this time he met with an accident after which he got an excuse not to go to work or earn, when he could because by this time his work did not involve any physical labour - he used his mind but he had broken his hand. Then he took the money that we had saved, nearly Rs.22,000 (£300) and returned to his village without paying his medical bills at the Sajjad Nursing Home where he was admitted. He never returned from there.

Our life continued in this fashion. Then in August of 1991 we had a big fight. My husband said that he wanted to sell the house and I said "no I will not let you sell it! where will I live?" I had lost my ability to tolerate and to ignore his bad habits. I never complained to anyone and he did what he felt like doing. "You encouraged me to progress at one point in life, let me form the union and register it, and if you want me to quit all that later then it is unacceptable to me. Now that I have progressed so much then why should you hold me back?" He did not like that.

Then he began making trouble and threatened to sell the house because he was unable to earn. We had a big fight, he became very angry and he beat me up and after that he filed a report against me in the Hanumanganj police station.

My daughter was not a bride even for one night

There is great social prejudice against gas affected people. It was difficult for me to find a husband for my daughter. Nobody wanted to marry her because she was gas affected so I had to marry her to a gas affected person. The man she is married to suffers from a serious skin ailment due to gas related medical complications.

My daughter was married before and divorced on the first day of her marriage. Her father found the groom from his village. My daughter was not a bride even for one night and her life was destroyed over the issue of dowry. They demanded the money from the first compensation and a car, when I was incapable of buying a cycle. Her father had abandoned us and now he was responsible for this. The wedding happened as planned but they demanded a car in dowry. When I could not give them a car they abandoned my daughter and never came back.

Then I sued them and fought the case for three years and made them return the dowry that I had already given them "give it back, you don't want to keep my daughter, return her dowry". I am a fighter. I won the case in the court. Then I got my daughter married.

Even Mansoor Ali, my second son, had trouble getting a non gas affected bride and is now married to a gas affected woman after much difficulty.

We already have one Bhopal, why make Gujarat into another?

The Government has collaborated with the company and the organisations do not want the Government to clean up the factory site. If the Government cleans up then it will not take responsibility, it will clean up and leave it at that and the organisations believe that only Dow should clean up and take its waste back to its country. If the Government cleans up, where will it dump the poison? Pritampura near Indore, so another Bhopal, they are also our brothers and sisters. The plastic inside the poisonous pond (Solar Evaporation Pond) of Bhopal tore after so many years and the waste in the ditch seeped into the water and then contaminated all the groundwater. Now we don't have clean drinking water and the people are forced to drink poison, after many years even Pritampur will be the same.

The Government was talking about incinerating it in Gujarat, they are our brothers and sisters too, so why should we do that? We already have one Bhopal that is destroyed and we make Gujarat into another Bhopal? Or we make Pritampur into another Bhopal? This policy of the Government is wrong. Why is the Government taking on the responsibility of such a big company, there must be some benefits, some bad intentions behind its hunger for the dollars of the company?

So we want only Dow to clean up and only Dow to take responsibility of Bhopal. If the Government cleans up the waste lying in the *godown* will it also clean up the waste under the soil? This is Dow's work, its responsibility and only Dow can clean up, its depth and spread, who will clean up the poisons in the ground? Only Dow, and Dow should take its waste back and take full responsibility. If the waste is cleaned by Dow then it will also have to take the responsibility of Bhopal and it is Dow's duty. We all want the clean up and we are demanding the clean up but we are demanding that Dow clean up. Why is the Government interfering? We will not let the Government clean up.

When the issue of clean up was brought up recently our organisations (in ICJB) opposed it. The Government brought in Ramky Company for the job and we also opposed that. Later they announced that it will be sent to Pritampur for land filling, we opposed that too. The Government has no business cleaning up the site it is the responsibility of the polluter. The people who have been affected by the pollution that the waste caused in the water should be compensated. Only Dow should clean up and take full responsibility.

I think the Government should decide on what should happen to the site once cleaned up. I personally would suggest work sheds to provide employment to gas victims instead of a museum. There was a girl who had plans to make the factory site into a museum and she also received a loan of Rs. 3 lakh. I feel that if something is made over the factory site it will be used more by couples who want to spend time in privacy, or by drunkards or rowdies. Something similar happened in the Lal Quila and the Government began charging an entrance fee. It will make me very unhappy if the Government starts charging a fee here.

If Bhopal gets justice then the whole world will get justice

Hajra Bee

Many people come to fight for the rights of gas victims, to share their pain and grief, to participate in their fight, this is their sympathy. They take the fight of Bhopal forward, amplify the voice of the victims, want to get justice for gas and water victims so these are the people who sympathise who come among us. If they are well educated they can make their money in some way, when they have so many degrees then they will not find it difficult to get a job. They have the sympathy and they want Bhopal to get justice. If Bhopal gets justice then the whole world will get justice. Then Dow will not dare to repeat another Bhopal like tragedy elsewhere or even start a company elsewhere. It is scared now, I feel its fear and it will be scared.

All the books that are written and all those who use the gas victims, all the organisations that work with us and the way we are fighting: I do not feel that we are being used. Because there is some gain somewhere through our stories. We ourselves are poor, all gas victims are poor, all those fighting are poor people. So I would not call them wrong, anyone who writes our stories or whoever captures our words either through a book or a video. I wish that my voice, maybe through the medium of a book or television or paper or a film, at least if it opens up the minds of other people, refreshes their memory and maybe that it kindles some sympathy and people from outside will join our voice and our voices will get amplified and our struggle and fight will get strengthened so that we don't accept defeat. I feel very good that we are getting strength.

Sadhna Karnik Pradhan

Bhopal Gas Peedit Sangharsh Sahayog Samiti
Bhopal Gas Victims' Struggle Support Committee

Sadhna Karnik is founder and convenor of *Bhopal Gas Peedit Sangharsh Sahayog Samiti*, a national solidarity organisation through which trades unions and other social movement organisations have supported work at the grassroots. She provides a connection between the local struggles of the survivors and the broader programme of the Left in Indian politics.

BGPSSS can be contacted through Sadhna Karnik A–108 Padmanabh Nagar, Post-Govindpura 462023 Bhopal sadhna_karnik@yahoo.com

Twenty five years of struggle

I am in a state of mixed emotions! I could not imagine the state of affairs after 25 years in Bhopal. I think my love and dedication for victims empowered me. When I left home 25 years back seeking answers to several questions about society I did not need to go to any university to know about the root cause of social problems. During Bhopal struggle the poor and sick victims with their very rich spirit have been my great teachers! Bhopal reveals the truth of the system as it has revealed to me!

I do not know how I survived the critical time of physical and mental harassment in Bhopal. But with my commitment I am still with the

Bhopal victims on the 25th anniversary. I wish to be always with them! I wish to donate my body after death for the cause of Bhopal victims.

Immediately after the gas disaster I reached Bhopal and was part of the steering committee of *Zehereeli Gas Kand Sangharsh Morcha* (Poisonous Gas Episode Struggle Front) constituted by political and non-political activists, writers, poets, intellectuals, artists, students, scientists etc in December 1984. The joint and united struggle of *Morcha* achieved extraordinary results on every front: relief, health, legal, scientific, rehabilitation; by developing a strong mass movement of Bhopal victims. *Morcha* also started people's health clinic *Jana Swasthya Kendra* (JSK) for detoxification of Bhopal victims.

Within one and a half years the joint movement divided due to political differences and the leaders deserted the JSK and the *Morcha* without any discussion. Feeling that a united effort of *Morcha* and JSK was needed for a strong movement of Bhopal victims I continued leading the *Morcha* and running the people's health clinic JSK. Although I was with the movement and not any particular persons or groups, I was targeted by all the groups in Bhopal.

After some time I found the *Morcha* being misused by some people so I decided to merge *Morcha* with a more effective broad forum *Gas Peedit Sangharsh Sahayog Samiti* (Gas Victims' Struggle Support Committee) a joint forum of trades unions, science organisations, NGOs, mass organisations, women, student and youth organisations working since 1988 on several litigations of Bhopal. Later JSK had to be closed due to lack of funds.

During these 25 years and up until now my efforts not only suffered severe economic crisis but I suffered physical and mental harassment and torture beyond imagination. The movement I developed met major setbacks several times. I restarted it every time. My experience says that the genuine efforts by a woman activist for betterment of society could be attacked not only by selfish individuals and organisations but even by people of wrong understandings in any political party.

But at any cost I had to be with Bhopal victims. In my most difficult times during these 25 years, apart from my family and gas victims I want to thank all those who have supported me to continue my efforts for Bhopal victims. On the occasion of the 25th anniversary of Bhopal I greet all people working for Bhopal victims around the world appealing for a joint struggle. *Gas Peedit Sangharsh Sahayog Samiti* has formed Bhopal Sahayog Trust and I appeal to all to support the gas victims' welfare centre we want to develop for Bhopal victims.

My experience as a woman in Bhopal for 25 years

Sadhna Karnik Pradhan

"*Le mashale chal pade hai Log mere goan ke Ab Andhera Jeet lenge log mere goan ke*"

"*Puch ti hai zopdi aur puch te hai khet bhi kab talak lutthe rahenge log mere goan ke*"..

"People of my town have started marching with torches in their hands Now they will defeat the darkness."

"The slums, the huts, the fields keep on asking until when will we keep on losing everything?"

For the last 25 years I have sung this song with victims of Bhopal in several demonstrations during our movement for justice. This has become like a theme song of Bhopal. Each time while singing this song I felt that my spirit has united with that of the hundreds of gas victims standing or sitting in front of me. With me they would start moving their heads and hands and start singing with me. For a while we would forget the surroundings and the pain. It seems as if time has slowed down in Bhopal looking at the pain and the suffering. My 25 years in Bhopal seem as if only few months have passed.

I think that the real age of Bhopals and Hiroshimas is not counted in years but it is to be measured in eras.

Has Bhopal stopped happening? Every year since the beginning of the disaster I have been asked this question innumerable times! Are the victims still sick and suffering in Bhopal? What has the company given to them? What is the Indian Government doing? What are you doing in Bhopal?

I think one needs to be connected to believe pain!

Bhopal is not only a disaster. Bhopal is a disastrous process! Several Bhopals actually start happening in the life of the victims after Bhopal has happened. Like several Hiroshimas actually started happening for the victims after Hiroshima had actually happened.

Bhopal disaster cannot be separated from the victims' movement and the victims' movement cannot be separated from the people who are participating in it!

The big leap of a small city girl

After completing my graduation in 1977 I was searching for the root cause of the problems of this society. I had many unanswered questions in my mind. I loved nature and wanted to save the environment but wondered if working in the education system would change society or was it something else? I was not getting real answers.

Though my family was educated, like any other family in India my family wanted that I should find a good earning job and settle down by marrying a suitable boy. I was searching for my unanswered question, but the pressure and harassment for marriage increased and my parents had to suffer which is common in India. So I took the biggest decision an Indian girl could take. I decided to leave my home at Indore before marriage.

In a backward State like Madhya Pradesh whose capital is Bhopal, majority of the population is illiterate, backward, belongs to suppressed and exploited castes, where most women are uneducated and still practice *pardha* (veil) system and there is no history of any reform movement, my small step of leaving a secure home before marriage was a 'big leap' for a small town Indian girl.

The law of change

I went to work in Kishore Bharati, an NGO working in alternative education in Hoshangabad. I started my work in a mobile library. Our group used to take books to villages several kilometres inside the deep woods using bicycles. We also participated in a study circle doing collective study of the book *Das Kapital* by Karl Marx. This study circle made me understand that the basis of this society is the Economic System which manifests itself into education system, environment policy, culture etc. The second big lesson Marx gave me was that world is changing continuously - in fact everything is changing except the 'law of change' in society.

I got answers to my unanswered questions, that working on the problems of society on humanitarian grounds alone will not affect the root cause of the problems. But collective efforts made in direction of change of system will change the society.

What I did not know was that in the near future, due to the exploitative economic system of a rich country like America and its collaboration with a third world country like India, the biggest environmental disaster of this era was about to shake the roots of the whole world. This was to bring a drastic change into the lives of millions of poor people of India and with them my life was also going for a big change.

Bhopal should not happen even to the enemy

I was at Kishore Bharati when Bhopal gas disaster happened and with the group I immediately started for Bhopal. Union Carbide Company and the Government of India were hiding facts. We immediately started the first people's survey.

"We would have been better dead than alive on the night of gas disaster". Deeply injured and shocked victims expressed themselves during the house to house survey we were doing. Each house had a tragic story of its own. They were suddenly forced to run at midnight, half dressed. Their home town had become a poisonous gas chamber. The pregnant women aborted and delivered babies in the middle of the road and the foetus and the new born babies were stampeded by the breathless running mob. The parents lost their children and had to find their dearest ones among the dead bodies. Mothers and babies died during feeding. The innocent citizens of Bhopal did not know what was their fault for which they were punished so severely? The question is still in the air in Bhopal.

In the steering committee of the *Zehereeli Gas Kand Sangharsh Morcha* there were 4 – 5 women in the beginning who soon left and I was the only woman leader left in the *Morcha*. Initially everybody participated in the relief work. But we knew that this flow of sympathy and humanitarian emotion would dry up soon and the real battle would start. "Rights are never served on a plate" we used to tell the victims. A long, continuous, organised mass movement is the only way.

The criminal company UCC and Indian Government did not give any information about character and treatment of the poisons. This was to diminish the magnitude and gravity of the disaster. Due to this there would not be any strong evidence in the court and the victims would get minimum compensation. With a weak case the guilty company could not be punished.

On the basis of scientific facts and first hand survey of the victims we made a comprehensive charter of demands for victims. The demands included short term demands of immediate relief: free milk, ration, medicines, relief money etc. and long term demands of compensation according to the magnitude and gravity of the disaster, rehabilitation, objective treatment of the poisons, writing 'MIC affected' on medical papers, scientific treatment of the residue of poisons lying in the MIC tank etc.

There were many achievements for the gas victims as a result of the *Morcha* which I was part of: free rations, milk, relief money, writing 'MIC patient' on medical papers, correction of claim forms, starting of 100 bed hospital and several polyclinics, interim relief, claim camps, work sheds in victim areas, work orders for victim women, employment loans, starting of *anganwadis* (mother and child care centres) for children below 14 years giving them milk and food, Public Interest Litigation in different courts of India and outside, gas *rahat* ITI for youth employment training, detoxification treatment etc.

People's Health Clinic: Jana Swasthya Kendra

The struggle in Bhopal could not be only in one dimension but it had to be multidimensional. In 1985 we decided to start People's Health Clinic

Jana Swasthya Kendra (JSK) in Bhopal. Union Carbide Corporation USA was suppressing facts about treatment of the poisons. The postmortem reports of the dead bodies of victims performed on the night of 2nd December 1984 showed the cherry red colour of the organs which was clear evidence of cyanide poisoning. But the criminal company UCC refused to acknowledge the presence of cyanide in the poisons that leaked that night.

A German doctor had brought few hundred doses of the cyanide detoxificant sodium thiosulphate (NATS) to Bhopal. State government provided them only to the rich and influential people in Bhopal. JSK developed a health card for Bhopal victims. In the health card we noted down the symptoms common among the gas victims. We recorded the relief in various symptoms before and after the course of injections of NATS. We also took urine samples of the victims before and after the course of NATS and got the urine examined at the forensic department of the government medical college. The higher than normal level of thiocyanate in urine was a clear indication of cyanide poisoning. This was evidence of poison on paper. The gas victims also felt relief after taking the course of NATS.

The ICMR (Indian Council for Medical Research) in 1985 also recommended NATS to benefit gas victims after their pilot project. JSK also performed various medical studies on women's health etc. In the beginning, *Morcha* opened the clinic on behalf of the movement inside UCIL premises where it was attacked by police. All the injections and other documents were seized and the voluntary doctors were arrested by the police. Later on this clinic was run in Kainchichola near UCIL in a rented house. The Government stopped us giving the NATS injections but JSK went to the Supreme Court. We could get NATS for victims only through court order.

Morcha and JSK demanded NATS to be manufactured in India and as a result Hyderabad IDPL (India Drug Pharmaceutical Limited) started manufacturing it. I remember one person from our group used to take the first injection from a fresh batch every time to see that the batch was not causing any reaction and was ok for the victims. JSK gave about

25,000 doses of NATS to about 5,000 gas victims. The JSK also did various health awareness campaigns among the victims. The information gathered from various studies done by the clinic was to formulate various demands for the struggle of the victims.

Nothing to lose but chains

The Bhopal disaster shows us that all barriers of caste and religion break during a crisis. On the night of 2nd December 1984 when humanity was bleeding in Bhopal the Hindus donated blood to the Muslims and Muslim to the Hindus. The Muslims set on pyre the dead bodies of Hindus and the Hindus buried Muslims. Both communities guarded each others' deserted houses on the night of the disaster. Victims in both the communities even looked after each others' children who were separated from their parents. During 1992 riots (after the Babri Masjid destruction), when the other parts of the city were burning, the gas victims guarded the area from communal riots. This communal harmony is a natural factor. But from time to time some fundamentalist organisations have tried to break this harmony for their narrow political interests. This has weakened the strength and unity of the victims' struggle. But in Bhopal the power of united struggle of all castes and religion has shown that only unity can take the struggle for justice forward.

• Work on war footing: It was nearly 24 hours a day work, we only rested for a few hours in the night and ate our food early in the morning and late at night. We contacted gas victims by going from area to area and making them aware of their rights and demands. We adopted several ways for mass contact: doing house to house calls, street corner meetings, distributing pamphlets, educative exhibitions, auto-rickshaw mic announcement and street plays etc. During the street plays the people donated us money to run the mass movement. The victims participated in hundreds and thousands during various agitations.

• A month long *dharna*: *Morcha* organised a nearly month long sit in (*dharna*) at Chief Minister Arjun Singh's residence in 1985. Hundreds of victims kept on coming and joining the sit in each day. Previously only 6 wards of Bhopal were declared gas affected. The victims from other areas and wards kept on coming and joining the movement and they informed us about the effect of the gas in their area. We kept on including them in the charter of demands and they were declared as gas affected. So in the beginning 6 wards were declared gas affected, then 10 wards then 16 after that 19 and the number reached 36 wards as gas affected in Bhopal. Later ICMR also declared 36 wards as gas affected by doing scientific study.

At the *dharna* at Arjun Singh's residence, the whole of Bhopal stood up to support the movement. Gas victim activists collected *rotis* from victims' houses and brought them to the *dharna* place to eat. The old city victims sent to us tons of breakfast. We were totally dependent on people's resources.

• *Rail chakka jam*: We decided to stop the trains as a form of protest – the *rail chakka jam*. The police wanted to arrest us before we could do it. So I and other women co-activists borrowed *burkha* from the Muslim gas victim women and wearing *burkhas* we walked in front of the police but they could not recognise us. In this way we reached the railway tracks and stopped the trains in protest.

• *Vallabh Bhavan* agitation: Thousands of victims signed a personal demand letter to the state government. We decided to hand over collectively the charter of demands to the state government at *Vallabh Bhavan* (secretariat). But nearly all of us were arrested in the middle of the night before the demonstration. Male police took us away at midnight without a ladies' escort. But the next day we found a huge demonstration of gas victims coming to our rescue and demanding our release from police control room. We were all released without any condition. Our huge demonstration marched towards the secretariat. The administration pre-planned to sabotage the movement by using criminal elements to throw stones from the back thus giving police a chance to *lathi* charge and make arrests. Strong criminal cases were imposed on us. We were all sent to jail for several days. The poor gas victims were also put inside the jail.

Apart from these actions there were several small and big agitations ie. *chakka* (wheels) jam, sit in, demonstrations in Delhi and in Bhopal, meeting state and central leaders.

The division of *Morcha*

During 1985 there was a two day meeting of the national campaign Committee on Bhopal Gas Disaster in Delhi. All the main leaders of the gas movement went to Delhi together to attend the meeting. I was the only woman leader left in *Morcha* so I was the only woman who went with the group to Delhi. On the first day the meeting concluded at about midnight. All gas victim leaders started departing. I told everybody that I did not have an address of any known person in Delhi and I had no place to go that night. I told them if they did not help me I would be left alone in Delhi at midnight. But not a single so-called progressive male leader of Bhopal helped me. All Bhopal leaders left one by one and I was left alone at midnight in Delhi without any shelter.

Anything could have happened to me that night. I think after this episode I should have anticipated the fate of the struggle, *Morcha* and the real faces of leaders of the Bhopal gas movement. In fact this was a way of warning of the destiny of what was going to happen in future in Bhopal.

Many important matters were not being discussed with me but I did not know that even the leaders would start leaving the movement and were planning to divide the movement without any prior information or discussion during 1986.

Suddenly one morning I was informed about a meeting of the steering committee of the *Morcha*. During the meeting a division of the *Morcha* was declared unilaterally. I was asked if I was with *Morcha* or with JSK. I was shocked because we were in the midst of a serious struggle for victims. A joint and united struggle was the need of the hour and any damage or division in it would severely damage the cause of the Bhopal victims.

Later on I could understand that the division was among two political groups. I did not want to be with any particular group or person. I only wanted the ongoing united struggle of gas victims to continue. I thought it was the worst time to divide the gas movement into *Morcha* and people's clinic JSK I wanted to continue to work with both *Morcha* and the JSK. I proposed a meeting of gas victim activists to take their opinion for a decision to stop division. But none of the fighting leaders was prepared to give me any attention.

I was going through a state of deep despair. At the time when Bhopal victims needed a strong joint movement the untimely division of the movement of the most sick and suffering victims in the world was most shocking.

The poor gas victims became my source of inspiration. Victims always wanted a joint struggle. In spite of their pain and sufferings they were ready to fight a long battle with great spirit. They faced big problems but had continued their struggle. The women woke up early in the morning, prepared food for the family, brought water from quite a distance, sent their children to school and their husband to find job for the day carrying a small Tiffin and came to meeting. The victim women still do it when they come to meetings and demonstration.

The division in the movement at the wrong time had weakened the struggle for justice of the victims. Within one and a half years of my coming to Bhopal I had to run the people's health clinic JSK and the *Morcha* all by myself. Both were deserted by the leaders as a result of the untimely division of joint mass movement of Bhopal.

Before the division of *Morcha* and JSK a hurried study was done in JSK which said that the NATS therapy was a psychological effect but ICMR had recommended NATS to Bhopal victims. During 1986, with the help of two voluntary doctors, I continued running the JSK alone in Kainchi Chola. In JSK we continued the sodium thiosulphate detoxification therapy. I used to fill the health cards of patients and take the urine samples before and after the injections to the forensic department of the government medical college. The relief to the victims was scientifically proved by increased thiocyanate level in urine samples and statements of patients recorded before and after injections in health cards which showed tremendous relief to the victims.

I remember a case of a Malwa express driver who was driving the train on the night of the disaster and was in a train on Bhopal railway station on that night. He came to JSK for treatment and told me that he had taken treatment in hospitals of Mumbai and Delhi but did not get relief. He had heard about the effect of NATS therapy and had searched JSK to get it. The JSK doctor was on leave. I had taken training to give intravenous NATS. I gave him 12 doses of NATS. After the course he told me that he slept for the first time after the tragedy. The victims would run after me while I walked along the road and ask me about the injections. Due to funding problems I had to shift the clinic to a free of charge room in West Railway Colony welfare centre.

Railway gas victims struggle

In 1986 West Railway Colony became one of the centres of gas victim struggle due to the presence of JSK. Railway colony is one of the worst affected areas. The flow of poisonous gas was towards the east, towards the railway station. In their effort to escape the poisonous gas, thousands of victims ran towards the railway station and bus stand to get out of Bhopal. Railway employees were present on duty on the night of the disaster and performed their duty amidst the poisonous gas cloud. Many railway workers had died doing their heroic duty. Hundreds of railway employees and their families are still seriously ill. We made a charter of demands for them and organised several demonstrations in front of their Delhi ministry and Bhopal control room. We met the railway minister with a group of railway women several times and were successful in getting work orders for railway employees' wives for stitching railway uniforms. Due to our struggle a work shed was built in West Railway Colony for the women gas victims in Railway colony.

During 1986 we started a Gas Victims Unemployed Youth Front for the unemployed youth gas victims and started agitation. During that time the railway coach factory in Bhopal was being constructed. We demanded reservation of jobs for gas victim youth in the railway coach factory. Due to the long agitation 150 young gas victims got employment.

March ahead, light will come

During 1986 while running JSK and *Morcha* one day I got information from a journalist that the UCC and Government of India were going to make an unjust out of court settlement in the Bhopal district court. Other NGOs were silent about the unjust settlement. I decided to start the struggle against the unjust settlement of a multinational from America, the most powerful country in the world, and the Indian state.

It was only one and a half years back I had come to Bhopal. I felt the responsibility to launch a joint struggle against the unjust settlement. When I think back I wonder, being new to a mass movement, where I got the strength to restart such a hard struggle. And I think about my strong commitment for poor victims as my power. When I started the struggle I felt as if I had to draw a dim light of hope from complete darkness. I decided to make a full effort. I wanted to put all my strength into the effort.

I remember that I called a meeting of patients of JSK against the settlement. Five patients attended the meeting. I explained to them about the unjust settlement. We all agreed to fight it. I called up a second meeting in JSK and about 19 patients attended the meeting and we decided to start an anti-settlement campaign in the gas affected areas.

I used to start the anti-settlement campaign early in the morning taking a handful of gram seeds and *jaggery* as breakfast. In scorching heat we walked from house to house and from area to area campaigning. A gas victim women from West Railway Colony was with me in the campaign and her railway quarter was my shelter for food. From morning till night for several weeks we continued the campaign.

In 1987 we organised an anti-unjust settlement convention in West Railway Colony and about 100 gas victims attended the convention. During the convention I invited some other leaders to join for joint movement against unjust settlement. We all decided to intervene in

Bhopal district court in the case of the out of court settlement on behalf of the *Morcha* and JSK. We put a joint petition on behalf of *Morcha* and JSK to stop the process of unjust settlement and ask for *lakhs* of *rupees* of interim relief to Bhopal victims from Union Carbide Corporation. This was historically the first petition for interim relief demanded from the criminal company UCC in Bhopal district court by the gas victims.

I used to take a group of gas victim women to the front of the district court every day until this case was decided. During the agitation against the unjust settlement I went to all gas victim leaders and NGOs to appeal to them to join the movement. But every other gas victim leader and NGO refused to participate in the joint movement. During 1987 I remember on the day of the final decision even the district court had a big lock on the front gate because the bar association had joined us in our call of "Bhopal closed" against the unjust settlement. Only the room of Justice Deo was open for the hearing. The final judgment was historic and brought great success for us. We had won the case. Justice Deo had accepted our petitions as interveners and in order to scupper the unjust settlement, ordered the American multinational Union Carbide Corporation to give *lakhs* of *rupees* of interim relief to Bhopal victims. This was a historic judgment by any standards. Historically for the first time a multinational was ordered to give interim relief to the victims before criminal liability was decided. This made the criminal case against UCC stronger. Justice has recognised our peaceful movement of victims against settlement.

My decision not to break the joint movement of JSK and struggle front proved right. We were welcomed by Bhopal victims.

Sidelined and tortured

I was silent on many issues whilst developing a united struggle against the unjust settlement between UCC and the Government of India in Bhopal district court. But during 1988 I started asking questions about the untimely division of the joint movement and the stance taken in our petition and some other applications (put in my name, and on behalf of the *Morcha* and JSK without informing me) in the High Court and the Supreme Court of India where UCC had gone to appeal in the unjust settlement case. Many activities were going on without my knowledge and even if I disagreed. When I protested, an episode of harassment and cruel efforts to take control over me crossed all human limits. Everything went beyond words. I was cruelly sidelined, physically and mentally tortured.

With serious differences of opinion to a stand taken in Supreme Court, I went to the Supreme Court, Delhi with a railway gas victim woman and filed an application with the registrar explaining the stand taken by my section of the *Morcha*. I remember we had spent our last penny on a court fee and did not have money to buy a ticket back to Bhopal or even go to Delhi station. One NGO in Delhi bought us our ticket back home. Instead of cooperating and supporting me in this most critical time I feel the other gas victim leaders tried to spread misinformation about me.

During 1987, one day I was cleaning JSK in the morning when suddenly a white Ambassador taxi stopped in front of the clinic door. Three very tall dark and red eyed persons came inside the clinic and stood at the door. I asked them who they were and asked them to sit down but they neither answered my question nor sat down. Suddenly I sensed danger in their unusual behaviour. They told me that they had come to take me for an outing. I realised that they had come to kidnap me. My brain rushed to face the situation and find an emergency solution. Suddenly I started talking to them non stop. I told them who I was and how I have devoted my life for the gas victims. I asked them about their religion and told them that like in their religion they fasted and during the fast they only think about their God, so like that I was devoting my life for the gas victims. I asked them about their family and if they need our clinic for any family member I will give them full support. Suddenly they stood up and asked my pardon. They told me that they had got wrong information about me. They asked me if I could treat their little daughter's eye problem and I said I will do it for sure with help of our clinic doctor.

During another incident in 1985 I was returning late at night with my co-activist. Suddenly an *auto* stopped and three persons with blood stained clothes came out and demanded money. My co-activist said we have two *rupees* take one *rupee* and let one be left for our fare. They said that they had just murdered a person and will not hesitate doing so again. Suddenly they saw me and asked me if I was with my co-activist. I said yes then he said politely to me that madam you came to our area for survey of gas victims, we know you. Then told my co-activist: "you are saved today because of her!"

Bhopal movement in Norway

During the year 1990-91 environmental ministers of 33 European Union countries held a meeting in Bergen, Norway. The people's representatives from all over the world gathered in protest against the policies of the rich countries and for sustainable development. I was invited by the people's forum as a third world woman representing Bhopal. We formed an international support group for Bhopal during the people's convention. We decided to do whole day fasting and then a die-in action blocking the front door of the hotel where the environmental ministers had their meeting. I coordinated the action and after the day-long fasting the Bhopal support group blocked the front door of hotel by lying down on the floor as dead victims of Bhopal. Our plan was that if police made any inquiry to us we will tell them our name is Bhopal and we have come from Bhopal. Then the police will arrest us. The Bergen police are infamous for releasing dangerous dogs on activists. But astonishingly the police came in our support saying that we support the demands of Bhopal victims. Bhopal support group was lying down as dead victims for more then four hours. Nobody could go in or come out of the hotel. Bhopal group demanded that I should address the conference of the environment ministers.

But America put a veto on this move. At last it was decided that the President of the ministers' conference would address and circulate an open letter from Bhopal on my behalf. In the letter we had demanded that Bhopal is the responsibility of the rich countries. We also demanded that the ministerial conference take concrete steps for Bhopal. Due to the movement's pressure I was invited inside the hotel and taken to the media room and shown live the President's address on my behalf and the circulation of the open letter I had brought from Bhopal to all ministers at the conference.

Later we met the senators of all the communist parties of Norway. After I came back, the Norwegian communist parties raised questions in the parliament. Bhopal support group held a sit-in at night in minus 7°C, for action to be taken by the Norwegian Government.

Demonstrations were done at the Indian Embassy by the Bhopal Support Group in Norway. We wanted the Indian Government to take action but it failed to do so.

Ongoing movement: employment campaign and medical fraud

During 1994 to 1996 we started a campaign to restart the closed worksheds for Bhopal victim women. We started public meetings at different stitching centres: JP Nagar, Chhola Mandir, Kapada Mill, Vijay Nagar, West Railway Colony etc. and put people's locks on them. We demanded the state government restart the worksheds and give employment to hundreds of gas victim women unemployed due to their closure. We also started movements against fraudulent claims, medical categorisation, registering new deaths due to poisonous gas and accepting children born to gas victim mothers as gas affected. Due to the stitching centre movement the Government handed over the worksheds to NGOs.

The Indian Council of Medical Research was doing research on all important health issues and the medical treatment of Bhopal victims. But UCC influenced the Government of India to suppress the results as effective research and its results would have made the legal case of Bhopal victims stronger. All vital findings of the research were suppressed and the future research stopped. Due to this since 1987 the line of treatment of victims could not be reviewed. We raised this issue during several agitations and in Supreme Court hearings.

The medical categorisation fraud is the biggest medical fraud in the history of mankind. To weaken the legal case against compensation claims and thus drastically reduce the compensation of the seriously ill Bhopal victims, UCC with support of the Indian Government sabotaged the categories under which the Bhopal victims could recieve medical examinations in support of their claims. During this categorisation fraud only about half of the victims were medically examined. X rays and other vital reports of the serious victims were certified as normal. Thus 98% of the serious victims were categorised in either 'A' or 'B' claim medical category which means either they are not affected or they were affected but now cured. This totally weakened the court case of the victims. Thousands of victims of claim medical category 'A' or 'B' have died due to serious diseases but their claim medical report shows them normal. Due to this fraud medical categorisation, the majority of serious victims got compensation equal to nothing. In our recent petition about to be filed in the High Court we have demanded a review of this fraud medical categorisation.

On the basis of the number of deaths on the night of the disaster, ICMR divided the gas affected area into 1, severely affected 2, medium affected area 3, less affected area. The reason for the highest number of deaths in severely affected areas was that they are surrounding the UCIL factory in Bhopal. The most deadly and heavy gases such as Cyanide were found in the atmosphere of severely affected areas even five days after the disaster. The severely affected area victims suffered from most serious diseases i.e. cancer, TB, lung fibrosis, kidney, heart attack, gynecological disorders etc.

Since the beginning, our organisation, unlike other NGOs, put emphasis on giving priority to severely affected area victims in all schemes of compensation distribution, medical treatment, rehabilitation etc. But due to lack of priority the victims are dying a neglected death without getting any benefit.

Compensation struggles

Some organisations demanded in supreme court a flat rate compensation to all i.e. same amount of compensation to serious, medium and less affected areas which resulted in same amount of compensation to serious victims about to die and to victims with minor effect of poisonous gas. This was not only highly unjust but no priority was given during compensation distribution to the severely affected victims who were about to die. Hundreds of victims died waiting for their due dates in the claim courts for hearing. The compensation amount demanded by these NGOs, Rs.50,000/- to all, was highly unjust , without any scientific basis and not according to the damage done to the life of the serious victims. Whatever is demanded from the Government, only one quarter is granted so that finally victims got compensation in negative because interim relief was cut from the final compensation and after paying bribes at various levels in court, and the money spent on their private treatment for several years, the victim got compensation equal to nothing. The poor illiterate victims did not have money or information to go to the appeal court. Even most of the appeals made were rejected. We demanded priority and higher compensation to severely affected area and victims during compensation distribution.

During 1994 the severely affected victims were dying without getting any compensation because no priority was given to seriously affected victims and areas and the long time taken for deciding cases by the claim courts of Bhopal. These courts had become centres of corruption. We started a letter to the Supreme Court campaign for organising a 'people's court' (*lok adalat*) in Bhopal for 'one day disposal' of the claim cases. About 5,000 letters were sent to the Supreme Court from several gas victims' centres. We formed five gas victim meeting places where regular meetings were held i.e. J.P.. Nagar, Vijay Nagar, Karod Widow Colony, Chholamandir and Railway Colony. Regular demonstrations were done at each centre and letters signed by gas victims were posted to the Supreme Court. Our demand of *lok adalat* was accepted and 'the peoples court' was held in Bhopal. Due to this, thousands of serious and needy victims got compensation in one day before they died.

Since 1988 BGPSSS (gas victims' struggle support committee) is playing a prominent role in drafting and fighting litigation in various courts of India and outside on important issues of Bhopal victims. Several cases are in process and others resulted in getting important court orders or directions ie. criminal case against UCC-Dow in Bhopal district court, environmental pollution case against UCC-Dow in the High Court, several other Public Interest Litigations (PILs) and case against Bhopal memorial hospital trust in Supreme Court, the PIL of 5 times compensation in High Court of M.P., environmental pollution case against Dow in US court etc. Due to these litigations the various courts have passed important orders, and directed to constitute high level committees ie. Supreme Court medical advisory and monitoring committees etc..

The Government of India was hiding the fact that about Rs 1,500 *crore* were lying with reserve bank as interest money from previous compensation money distributed with public agitation, signature campaign and petition in Supreme Court. We won the case and about six *lakh* Bhopal victims got second time pro rata compensation of about Rs 25,000/-each.

During unjust settlement the Indian Government put to the Supreme Court a figure of about one *lakh* two thousand as affected. However it distributed the compensation received among six *lakh* seventy two thousand victims. Thus each victim got only one fifth of the compensation they were entitled to.

We did several demonstrations and representations on the issue. Our recent petition has been signed by one *lakh* victims demanding five times compensation to the victims and asking for an overall review of the unjust settlement in 1989. This PIL is about to be filed in the High Court.

Bhopal Memorial Hospital Trust – or mistrust

Bhopal Memorial Hospital Trust was constituted by UCC to save its assets from confiscation. The trust and the hospital run by it had several incidents of corruption and irregularities. The trust and the hospital were more interested in treatment of rich private patients than the poor gas victims. The serious gas victims were given dates long ahead for operations and were discouraged from being admitted into the hospital. The staff of doctors and nurses are highly dissatisfied due to the attitude of the management. In 2004–2005 the doctors started total strike of the hospital staff. We joined in their struggle. The strike continued for 40 days, but the management was autocratic and refused to hold talks with the doctors. The gas victims were dying without treatment. The chairman of the trust who is a retired Supreme Court judge used his influence and meetings with the Prime Minister of India to put a notice in the hospital that the staff will be thrown out of the hospital with the Prime Minister's order. When I was informed about the strike I immediately rushed to the hospital. Sensing the emergency I immediately took a decision to fast unto death. After starting my hunger strike, within hours the management called the doctors to negotiate and a compromise was reached. The hospital was reopened for the gas victims.

Pollution, poverty and political power

The UCC had been dumping tonnes of hazardous poisonous waste inside and near their factory premises in Bhopal. These poisons have filtered into the ground poisoning the underground drinking water sources. Some agencies declared a few areas of Bhopal as poisonous water affected. But we found that even the area adjoining to the dumped poisons site is not declared as poisonous water affected. So since 2005 we have been demanding a scientific, in-depth and in breadth investigation in about one mile radius of the dumping sites of Bhopal. The poisonous water affected victims should be registered for medical treatment and claims.

As a majority of the victims suffering with serious diseases are daily wage workers they eat only one meal a day. Sometimes they take the stale food of the previous day. The majority of Bhopal victims come under the category of Below Poverty Line (BPL).

Bhopal victims depend on BPL ration cards for food for their family. Due to these BPL cards they get subsidised ration of a few kilos rice, wheat etc. But hundreds of poor victims do not have these ration cards. The Supreme Court of India has given clear orders not to stop making these cards, but for years in Bhopal the Below Poverty Line survey was not being done and thus these ration cards were not made. During 2004–2005 we decided to launch an agitation for a demand to immediately start a Below Poverty Line survey and to give victims these ration cards. Due to several demonstrations on the issue the Below Poverty Line survey was restarted in Bhopal and thousands of victims got their Below Poverty Line ration cards.

A majority of the victims are unorganised sector workers earning daily wage: *beedi* rollers, construction workers, domestic workers etc.. We organised several demonstrations from 1996 until 2009 with trades unions, for implementation of welfare schemes for them.

Apart from working on the main issues of victims we carried out several other activities in several areas. Fifty literacy groups were formed, rehabilitation training programmes, awareness campaigns among unorganised workers for welfare schemes and forming self help groups etc.

The Bhopal disaster happened due to the double standards of rich countries towards people of poor countries. Various governments play into the hands of big industrial and corporate houses which in turn play with lives of innocent people. Hazardous industries use dangerous chemicals and occupational hazards are the results of the hunger for profit of multinationals, the cost of which is in people's lives. But sustainable development is a political fight for pro-people policies. Anti-people policies cannot be changed without changing the anti-people political system.

The political impact of the Bhopal movement

There is no doubt that a large number of women participate in the Bhopal movement. But Bhopal has not been a platform for particular women's issues, or their political and general demands, so the mobilisation of large numbers of women victims following their male leaders in various organisations does not mean that Bhopal is a women's movement. However, the Bhopal movement has contributed to women gas victims becoming more socially active.

Bhopal gas disaster movement is mainly an issue-based movement. The victims are mobilised mainly for economic demands or compensation. But 25 years is a long time to judge its political impact. When we look back to see this in perspective we find that immediately after the disaster the victims elected the same party candidates in Bhopal and M.P. who were responsible for the gas disaster. Since then either corrupt or communal forces have been dominating Bhopal's political scene.

Even the gas victims who have been regularly coming on demonstrations for the last 25 years and gas victim activists in various progressive organisations are voters or supporters of corrupt or communal political forces.

Though the gas victim NGOs in Bhopal are prepared to take the support of all political parties they oppose all political parties. Due to this political stand of the non political NGOs in Bhopal, even after a 25-year-long progressive struggle the victims who are followers of various NGOs have to elect communal and corrupt political forces and an alternative pro-people political force could not be developed. Left parties have played a very important role in raising various issues in parliament and outside, especially at the time of the interim relief movement.

Due to a lack of understanding of the need for a united, broad movement, the Bhopal movement could not have an impact on the use of hazardous chemicals, pollution, occupational health and safety of workers in different industries in Bhopal and outside. For the last 25 years the industries using unsafe technology and hazardous chemicals around the world seem to ignore Bhopal's lessons.

If we need an organisation for gas victims' issue-based struggle in Bhopal then perhaps we need a bigger and more powerful political organisation for a political struggle against anti-people policies working in favour of rich countries like America which results in many 'Bhopals'. The goal of sustainable development cannot be achieved without changing the political system responsible for making unsustainable policies of industrialisation.

Bhopal, part of a global struggle

Although all over the world due to the impact of globalisation, communalism, regionalism, casteism and consumerism have diluted the progressive political movement, still we find that the Bhopal movement could not develop any alternative progressive political force for the victims.

Struggle for various issues of victims in Bhopal will continue but Bhopal cannot remain isolated from the broad struggle going on at global level against globalisation and anti-people policies of the rich countries like America which are severely damaging the environment and life of poor people of third world countries like India. Unity of Bhopal with a global struggle for pro-people policy will really start to affect the root cause of the Bhopal disaster and empower the victims to solve the various issues to ensure that there are NO MORE BHOPALS.

Om Wati Bai

International Campaign for Justice in Bhopal

Om Wati Bai is one of the committed rank-and-file activists who is not specifically affiliated to any group. Tending to be involved in ICJB, she has also supported BGPMUS and others. Intelligent and reflective, she nonetheless has no leadership aspirations.

Only the permanent workers had effective unions

I used to live near the *subzi mandi* (vegetable market) Ram Mandir in a rented house and I moved into J.P. Nagar next to the factory in 1975. A lot of people were settling here, my neighbour's mother told us about it and she helped us build our hut. Then in May or June 1984 *pattas* (land titles) were given out and in the middle of August we got an electricity connection for one bulb. There were around 800 to 900 families by 1984.

After the disaster a lot of people sold their plots and settled in new slums elsewhere. More than half of the people of J.P. Nagar have gone to areas like Chouksey Nagar. A lot of people sold whole lots of land and moved onto footpaths in Indira Nagar.

My husband was a contract labourer. He would hear about vacancies through notices on the factory gates. People would gather at the gates of the factory, some of them would be selected and taken inside for some tests (height, eyesight etc) and then hired. There was a union which he was a member of but it was ineffective and he was never interested in their affairs. Only the permanent workers had effective unions. As a result, when he was hired, the permanent workers got hats, face masks, gloves, socks, protective clothes etc. and the contract labourers just got face masks.

Even though we were aware of such practices, when they built the Union Carbide factory nobody imagined that such a thing as the gas leak would happen. The farmers sold their land for the factory to be built because they were getting good money for it, and people did not oppose because they did not know what was coming.

I remember all my old neighbours from those days. Only a few young people have moved out after they got married but all other families are still around.

There were no leaders at first

The day after the disaster the people were really angry. There were furious protests outside the factory; people were threatening to burn the factory down. That was when someone spread the rumour of a second gas leak and caused the crowd to disperse. There were no leaders as such because everyone had to take charge of their own lives. There were a few people from the *mohalla* (neighbourhood committees) like *chakki wale* (flour mill worker) Ismail, *beedi wale* Babu Dada, Ram Singh (who is no more), Purshottam Singh etc. These were the ones who took the initiative and would gather people for demonstrations. There was no compulsion to join them, we would go if time permitted. At that time my mother and my sister also lived with us, so there were three families and at least one person would join the protests.

Then Sathyu, Sadhna, Alok *bhai* and all the big shots were there. The leaders of the *Sangharsh* came to us and all the slum dwellers and we did not have any means of livelihood, and they opened their office at the *aloo* factory, and then they opened the hospital at Kainchi Chola, and then they used to tell us to go here and there, and I use to go with them, so I started to get to know them. They went ahead and we just followed them.

People still vote with the hope that the next government will be more sensitive

There wasn't much help from the Government. They distributed milk and bread after 2 days. It would arrive at 5 o'clock everyday, everybody would get a mug and we got wheat and cooking oil for 3 months through the public distribution system. The food came in big vessels and we did not know if it was being given out by Hindus or Muslims. Before that time I personally would never drink water from the home of a Muslim or ever eat with them but this changed after the disaster. We shared what was available, everyone the same. That is when I can say that the differences between Hindus and Muslims vanished. But even today, if there is a call for support a Muslim *basti* will have a better response than a Hindu *basti*.

My priority at that time was to punish the offenders and get compensation. Now I would also put more emphasis on the Government because it is just as much their fault that all this has come this far. It is the Government's responsibility if they permit such factories but if after that something goes wrong it is the company's responsibility. Then I feel that the company is more responsible because they were careless and allowed such a big disaster to happen. People still vote with the hope that the next government will be more sensitive to their issues.

When the Government made the settlement with Union Carbide we were given Rs 200 per month interim relief money for about two years. One time I received only Rs 200 over a period of 3 months instead of Rs 600. When I asked them they told me that the relief money was stopped for a while because the Government had to divert the money to earthquake victims in Gujarat. I did not have a problem with that because it was also for a similar disaster. Then when we got our second compensation, the money we had received from the interim relief was deducted. I had already received Rs 12,000 and that was deducted from Rs 25,000. I used my compensation money to make my house.

I cannot think of any punishment suitable for Warren Anderson. The only thing that would give me peace is an assurance that such things will never happen in the future. Most people are still with the movement because they want to punish Dow. A lot of them also want a final decision on their issues. Who does not expect compensation? Everybody looks forward to getting money, especially when they are poor. However, I think if the people of Bhopal had to choose between compensation and punishing Dow they would choose compensation. People do not want to think about the future of their children. But if Dow gets away with what Union Carbide did to Bhopal, there will be nothing to stop them doing the same thing again. For me the priority should be that we do not make such mistakes in the future for which our children have to pay. My own grandchildren continue to suffer, two generations after the gas leak.

We lose if companies come and we lose if they don't

These companies are harmful so they should not come to India, but if they don't then India will not progress. So we lose if they come and we lose if they don't! The farmers managed their crops very well before these pesticides came and now the crops are safe from pests but the farmer is also paying a price for it.

I keep informed about what is going on in India. My source of information was news reports on the radio. I have had a 2 cell transistor radio since 1975 and that has been the source of information for me. I can't read so the most effective medium is voice; word of mouth or if someone reads it out to us.

I have been protesting for 25 years. I have been to Delhi 4 times and once to Bombay. I have been to numerous demonstrations in Bhopal: marches, rallies, *dharnas*, protests at the Chief Minister's Office. The survivors' movement will always behave with humanity and try all the peaceful means possible: we will go hungry; we will even break the law. We will protest peacefully but if we are not listened to then we will resort to violence. This will be the last measure. There is a just violence in which there is no place for injustice and if there is injustice in the violence then it will spoil the chances of victory.

The groups are fighting their own battles, but the war is the same

I never understood why we have all these different organisations all supposedly fighting for the same thing. I would join anybody who asked me to join them. I would never ask about the politics of the organisation or the individuals who lead it. I would join people if they would fetch me and drop me back. I used to go to the BGPMUS rallies if they organised the travel. However they make you donate a minimum sum of Rs 5 for everything they do to cover costs and I can't afford that. I would continue to go if I could see some hope of progress but if I don't see any prospect then I don't continue. The last time I went was to fill up the 5 times more compensation forms.

I never really paid any attention to how decisions are made in the movement. I always believed in following the leadership. I have just started attending these meetings where things are discussed and people give opinions. Sometimes when you want to speak someone interrupts you or gets rude with you, so I generally feel it's better not to speak. I do not want to take on so much responsibility. I am illiterate, what am I going to do with all that information? It is true that Rasheeda Appa is illiterate too and she is a leader. Appa shows that when a woman is leading and gets support she can do so much more. A lot can be done without education. The young people who are joining the fight now, many of them are at school or have some education. It would be wonderful to see more young people join.

If all these organisations work together the fight will certainly be stronger. But they do not get along because people want credit in their name for work that everyone has done. However, the fight would have not come so far without these educated people who have supported us. The men are in the forefront because they have raised these issues for the people and there should be no problem following their lead. Our effort should be to encourage this and help them take it forward. If a woman had taken the lead then she would have been in the forefront and we would help them. I trust all the leaders in the movement: everyone. Nobody would fight if there was no trust. So I trust that if they take a particular step it must be after some thought. I would not say anything bad about any leader because they are all fighting their own battles, but the war is the same.

Razia Bee and Ruksana Bee

Bhopal Gas Peedit Mahila Stationery Karmchari Morcha
Bhopal Gas Affected Women's Stationery Workers' Front

Rasheeda Bee and Champa Devi Shukla were elected leaders of the Women Stationery Industries Union in the economic rehabilitation work sheds. They led the fight to transfer the work sheds from state industries (Madhya Pradesh run small scale craft production) to government press central government owned production). Having achieved this the fight focused on obtaining permanent status for the workers. At the time of the interview a court case was pending concerning the right of the stationery workers to permanent employment. The case was won in December 2008.

In 2004, Rasheeda Bee and Champa Devi jointly won the Goldman Prize for environmental campaigning, for the work on behalf of the victims of the Bhopal disaster. This led to a split in the union between those who supported their acceptance of the prize and those who believed it had no benefit to the union which had elected them. Razia Bee and Ruksana Bee are part of the breakaway union. This interview was conducted with both of them together and the resultant piece is an amalgam of their responses.

Rasheeda Bee was our leader

After the gas tragedy, a man came and told us that we would get work in the stationery workshops. So we left school and they gave us training for one month, then we were sitting at home. Then one day a boy came and said if we wanted work we could come and work here. The work was from Allahabad bank. There was so much work we would be here from 8 am until 8 in the evening. We had so much work we used to get boils in our hands, cutting, pasting etc. Our union was formed when we were shifted to this place. Our union used to fight for us, we had a lot of demands. We made Rasheeda Bee our leader. She was the eldest, and the most sensible, and we all were very young, and not up to it. So that is why we selected her. We used to save a little money, and then give it to the union to fight our struggle until we were made part of the government press. She used to consult the educated people, and then have meetings with them, and then she used to do the work. After the completion of work, she used to come and tell us that this is done, and consult us and we used to agree to it. If we used to say yes, then she used to do it, otherwise not.

We would do a lot of demonstrations, rallies, *dharnas* etc. We had a protest at *Vallabh Bhavan* (secretariat) for 27 days. We were in state industries, and we protested so that we could be put in the government press, which would mean better pay and conditions. In the nights we used to have torch-lit rallies to the chief ministers place, and to *Vallabh Bhavan*. We were getting worried whether they would do it or not. Then we did the *padyatra* to Delhi and we succeeded.

To start with there were fights at home about this. Before I was married my father used to shout at me, saying, "Why are you behaving like this? We will get defamed and you will not be able to get married because of it." When I went on *padyatra* to Delhi my family shouted at me. They would say "a young girl is going out, who will look after her, what will people say?" So we used to have fights at home.

Once we were in government press, our issue was to make us permanent. We did not want contractual work, because we had to work a

Razia Bee & Ruksana Bee

lot and received very little money. We had to make 160 files, and there was a lot of labour in that and the pay was very low, so we wanted to become permanent, so we did not need to do so much labour for so little money. This was what we have been fighting for. There was a lot of work to do, first get the paper, then make the glue, stick to the craft paper, cut the paper, then dry the paper, walk on the paper to stick the paper. A full day we used to work in the sun and then in the evening collect the finished thing and keep it inside. It was a lot of work, but the rewards were nothing.

We never worked with Jabbar's union. Our fight we have fought by ourselves. Rasheeda Bee never wanted any other banner with her and wanted to fight alone and with her own brains.

Rasheeda Bee was our leader. We used to be afraid of her, even looking at her eyes we would be afraid of her. We would fear her anger and if she said to us sit, we used to sit, and if she said stand, then we used to stand up. Now everybody is angry with her because she left us and joined Sathyu's organisation. Now she will not come back because Sathyu is getting some advantage from her and nobody is going to leave a thing from which there is some benefit.

She got some things wrong too, like she was supposed to fight our case for permanent employment in the labour court here at Bhopal, but she went to Jabalpur [High court] and it was sent back to Bhopal. They had written to us that we had lost the case, and it has to be fought in the labour court, and then in labour court we won the case, and the appeal we also won, but in Jabalpur we lost the case.

Rasheeda Bee became famous. People started to know her and the lawyers she used to consult.

As the fight goes on our belief in God will go on increasing

We learned a lot from the struggles of these years. A lot of changes have come about. At first we never came out of the house and were involved in the daily activities of the home. We had never even seen the roads of Bhopal, and only when we came to fight the struggle we saw everything. Previously we just used to go to school and then back home.

My father never used to give me permission to go out and always used to say that when school closed we would all sit at home. Because of the struggle we got an opportunity to come out of the house. First I did not even know what a demonstration was, or a campaign or movement. We didn't know what it was like doing a job, or being in a union. Then we saw all these things.

My faith in God has increased since the gas leak, not diminished. The gas leaked. It was the company that had kept the poisonous gas which leaked, what is the fault of God in that? Had God planned that so many people had to die in the tragedy? Or so many would get employment, and then would be destined to die? They got both these things, employment and death. Our fight goes on and as the fight goes on our belief in God will go on increasing.

Padyatra to Delhi

We got the idea of doing a *padyatra* to Delhi from a big chairman from some political party. We can't remember his name but he came from Delhi and suggested the *padyatra*. So we talked to all the women, and they agreed to go for the *padyatra*. I thought that if I joined it I might get a good, permanent job. Also Rasheeda Bee, our leader, would put pressure on us. She said if we did not join the *padyatra* and did not do what she said, then we would not be able to come and work here any more. So out of fear we did it. When the day came, people put garlands on us, and then in the evening we started the *padyatra*.

We had lots of vigour, hopes, anxieties. We used to walk a full day in the sun, and if one woman got sick, then she would go and sit in the van. People in Delhi would ask where this organisation is from. When Rasheeda Bee was on hunger strike in Delhi, women would go from Bhopal in groups of five or six, the women of Bhopal gave her a lot of support. When we went on rallies, the fun and vigour that we had is something unique. Now, because of the effects of the gas, we get tired easily. On Monday we had a rally from Roshanpura to the Minister's house and we got so tired, all our legs and everything was in pain. Now we have become weak due to the gas, but whenever we do a rally or demonstration then we put life into it, and do it with gusto and fun.

To fight for the gas victims she should not have left us

After she started working with Sathyu *bhai*, she did not concentrate so much on her work with us, and most of the time she used to work there with him. He started taking her to London and America. Going abroad is good but she should think about the people who are left behind. If she thinks about others it is good. She has left behind the comrades who were with her in the beginning. Now, going to other places she should think about us too. The time she spent on our demands is much less than before. If she had not joined with Sathyu we would have won the case by now. She was with our organisation and used to fight and Sathyu used to come for the protests sometimes. But we did not know that he would also raise these other issues. And beside his issues, our issues used to have less weight. So therefore we did not like this. If you want to raise any issue, raise it in your own organisation, not with ours.

Then the dollars came [the Goldman Prize] and they tried to hide it from us. Sulakshna had told us that the money came in our organisation's name, and we should use it to fight our case. They tell that the organisation is so poor, and cannot fight its own case. People from abroad gave us money and help but Rasheeda and Champa Devi never allowed anything to come to us. So in anger we separated from them and formed a separate union. Twenty women went with them and forty are with us.

They have put some money into Chingari Trust and they do something for the handicapped people, but some money they eat. Now the money is coming in so she must also be spending on herself. From here she gets Rs 2,000 and from Chingari also she must be getting some money, so she is in a better situation than we are. We were losing out before, and now we are also losing.

If they had given money to the union then they would have sorted out our case by now, and now everything is pending in the court and they ask for money, sometimes Rs 100, or Rs 200 and from where can we get that money, when we get so little? The lawyers are also asking for money to fight the case and now that there are two organisations it is more difficult to get the work done. She is not paying any attention to the case, whether the lawyer is good or not, and if it is not a capable judge, then we will lose the case.

Then we discovered that that the union's registration had got cancelled. Rasheeda Bee used to deal with the registration, we just used to collect the money and give it to her. First she got the registration done but then she neglected to do it and for the last thirteen years she had not been keeping up the registration of the union. So all the women in the union had a fight with her, because she had not been giving us any account of what was happening. The registration was cancelled, and the union also got cancelled. We came to know this from the union office that this union is cancelled and you all have to make a new union, so we made a new union: the *Bhopal Gas Peedit Mahila Stationery Karmchari Morcha*. The *Morcha* is registered, but the *Sangh* has been cancelled.

To fight for the gas victims she should not have left us. We also would have fought for the gas victims. She left our issue and started fighting for other issues. Once upon a time Rasheeda Bee was something and now what has she become?

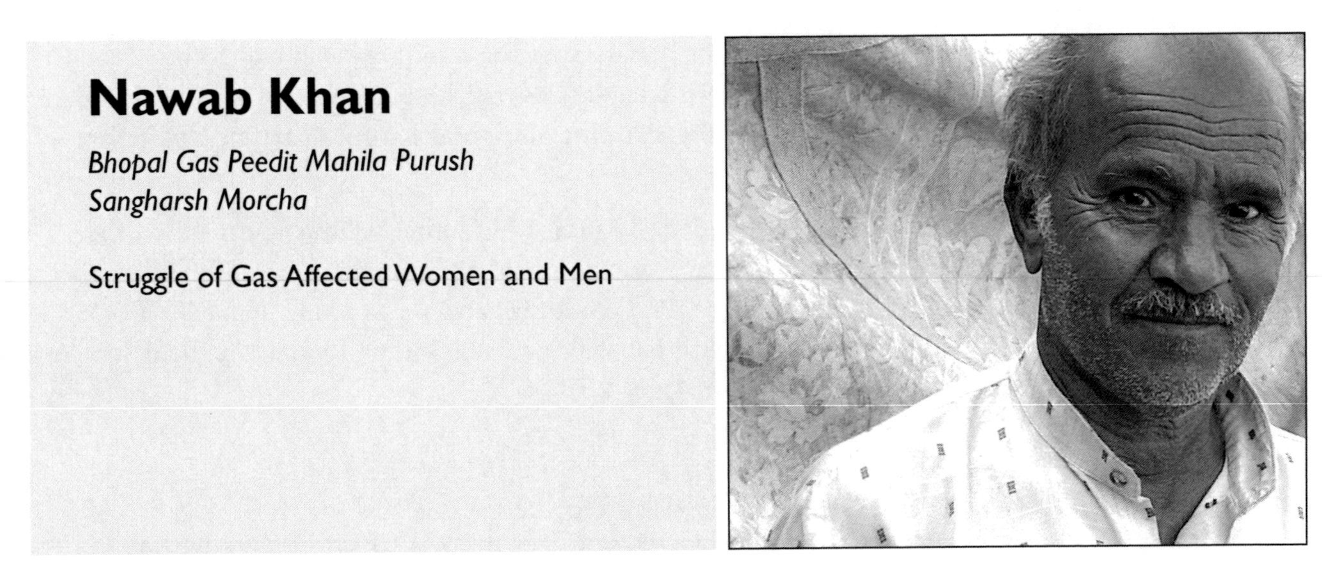

Nawab Khan

Bhopal Gas Peedit Mahila Purush Sangharsh Morcha

Struggle of Gas Affected Women and Men

Nawab Khan joined Syed Irfan when he formed *Bhopal Gas Peedit Mahila Purush Sangharsh Morcha*, and he continues to be active in ICJB. A journalist fluent in many languages, although amongst the leadership, his primary focus is in mobilising amongst the rank and file.

High time I joined the fight

On the night of the disaster my nephew told me about the gas leak. When I had been in Bombay I had witnessed an oil well accident and I assumed that this was a similar gas accident and that the whole place would catch fire. So we ran along with our children and reached New Market to a friend's house. My wife died in 1989 due to a prolonged gas related ailment. All six of our children faced the gas leak. I lost my eldest son in 1991 at 18 years old, to TB. He was a plumber by profession and he had started supporting the family with his income.

I was a tailor by profession and owned a shop in New Market. I was doing very well in business but after the gas leak I began making a loss because I was in the hospital for a long time, I had to pay rent despite no income, so I finally wound the business up. Around 1985 I had closed down my tailoring shop and I would earn my living by selling fruits.

When I was admitted to hospital I met other victims and realised the intensity of the issue. I was aware of the campaign because my wife used to attend Jabbar *bhai's* meetings and she kept me updated. When my wife and son died I felt that it was high time I joined the fight so I joined Jabbar *bhai* in 1986.

Bhopal: a fish bone in the government's throat

By this time I had also started writing for newspapers for a living since I have a good command of several languages including Hindi, Arabic, Urdu and Farsi. During those days I would write in a paper called *Saptahik Sundaram Bhopal*. Then I started my own weekly newspaper called *Sunehra Bhopal* (Golden Bhopal) in 1986 because I felt that the other papers were not giving enough coverage to the Bhopal issue. I had written a piece in the editorial in which I said "*Sarkaar ki halaak mein yeh kata kis taraah uljh jaayegi na nigaal sakte aur na bahar phek sakte*" "The Bhopal issue is going to be like a fish bone in the government's throat, that it cannot swallow or vomit" and that's what is happening. The then Chief Minister, Vora, called me and asked me to explain this statement and I just stood by it saying that it is the truth.

I spent 5 years as a member of BGPMUS without any responsibilities or post. I was friends with Irfan *bhai*. He was the first one to report the gas leak in Talaiya Thana. I got very close to Jabbar and Irfan *bhai*, but Irfan *bhai* and Jabbar *bhai* did not get along for some reason and separated. Irfan *bhai* told me how he was upset that he was given only clerical work by Jabbar, he was not allowed to do field work which he was interested in. Nobody was allowed to make any progress in that organisation so we should start our own group where we will fight for the poor. And we started the *Mahila-Purush Sangharsh Morcha* [men-women campaign front]. When Irfan *bhai* quit 15 of us left with him.

We did a few demonstrations and *dharnas* under this banner. Rasheeda Appa and Sathyu also knew about our work and when they formed ICJB they proposed a coalition because only by working together could we achieve success. We were unsure whether we could manage our organisation on donations alone but they did not believe in taking donations from gas victims because they had other sources of funding. Irfan and Sathyu were familiar with each other and they also held the opinion that a united force should be formed. So ICJB was formed and Rasheeda Bee was a part of it too.

The demands and issues raised in ICJB are what were discussed amongst all the groups and then put forward under the name of ICJB. Our organisation along with the coalition has been working only for the past 5 to 6 years. ICJB coalition has fought a lot of fights together and also won them.

I would not discredit Jabbar *bhai's* contribution to the fight. He has made the movement what it is and when the history of Bhopal is written he cannot be forgotten.

My organisation and the three organisations that we work with will never be invited to the table by the Government because the Government does not want any trouble, it prefers to collaborate with organisations that keep silent on its policies. I am still to find an honest officer or politician.

I have never seen myself as a leader

Before the gas leak I was a part of the Congress Party. I did not hold any particular post in the Congress, I was a grass root level worker and my role was to gather people when political leaders were visiting. I left Congress after 1984 because I was unhappy with the way they worked. The leaders made us run about and use us for their benefit or work, but when it came to helping us, like in the case of the gas leak they did not.

I have never seen myself as a leader. In the *basti* people come to me seeking advice on gas disaster related issues or other issues. I prefer working with the neighbours and not lead anyone. The problem with helping people in a *basti* is also that if you fail to help even one person your reputation could be tarnished. So I prefer not to get involved, I just direct them or take them to the meetings.

I am very happy with the way decisions are taken in ICJB because every one is consulted. Sometimes because of circumstances Rachna, Sathyu or Appa make decisions without consulting us but there has been no instance when we have disagreed with such decisions. Members have that power, even if I have to take some decision and am unable to consult anyone I am allowed to do that. That is the reason why we are all still together.

People affected by toxics

The demands have not changed but have been contorted by the Government. When we raised the issue of water contamination, the people would not trust us. Then Greenpeace tested the water and found poisons, and that did not change the situation much either but when the state Pollution Control Board released results and the Supreme Court ordered the Government to supply clean water people began to trust us. To deal with it we have to educate people and try to destroy the roots of such problems.

Since water contamination issues were raised, instead of saying water victims and gas victims they started addressing the people as people affected by toxics, because both groups are people affected by poisons and toxics. But still gas survivors have different issues and hence fewer gas survivors come and join in the struggle saying that we don't raise their issues. 25% are gas survivor's demands and 75% water survivor's demands. Whereas all issues came up because of gas survivors and their struggle they are being addressed as one now.

Egg *parathas* tomorrow but today dry *roti*

The biggest mistake the Government made was by accepting the settlement. Since then both the state and central governments have treated the victims badly. The settlement was inadequate because the Government figures were wrong. The figure for those injured was 1 *lakh* 5,000 and the dead were 1,500 and based on that each injured victim deserved

Nawab Khan

Rs 75,000 and the bereaved deserved Rs. 5 *lakh*. But the children who were earlier excluded had to be included in the claim list and many more had to be counted in. And hence 1 *lakh* 72,000 people were compensated. Otherwise each gas survivor would have been given Rs. 1 *lakh* 25,000 as compensation. But the numbers kept increasing. Later there was a lot of public pressure on the political representatives to have more wards included and the Government included the 36 wards in order to attract votes from those wards in the elections. But the settlement was not revised so the same amount had to be divided amongst more people.

It's the Government's fault. Why did they have to reduce the number and make it 1 *lakh* 5,000? If one is making a settlement for 3 – 5 *lakh* people then the appropriate amount should be paid by Union Carbide. But the Government made the settlement on a lower calculation, indicating that they must have taken some money under the table without which such a deal could not have been made.

Then another of the major issues in the movement was the second compensation. When the second compensation came through and we were given Rs.25,000 as compensation money, Jabbar announced to the people that we should not take that money but build pressure to get more. He wanted to push for Rs.1.5 *lakh*.

It is true that Rs. 25,000 was not sufficient, and from that amount the interim relief money that people had received was deducted, and perhaps we should have followed Jabbar's instructions. But like me, many were not in a position to decline it. At that time and in that situation Rs.25,000 looked like a lot of money, even if we had got Rs.10,000 we would have thought that it was a lot. "I may be given egg *parathas* to eat tomorrow but today even the dry *roti* that I am provided is good for me." This was supposed to be an interim relief amount and not the whole money due. So as interim relief, people had accepted it. Gas survivors were hence kept in the dark. It was easy to convince the survivors because they has already been given an interim relief amount of Rs.200 earlier and people believed this to be the same.

Justice for Bhopal could be different for everybody. To me employment is the priority then medical care and rehabilitation. Victims expect monetary compensation but we have seen people make a lot of money and they have nothing now. I am a gas victim and own 6 diaries of claims but I cannot even lend Rs 1,000 to anyone. If my children or even if I get employment I will be much better off.

The compensation money is misused by just a small minority of the victims who might have used it up in gambling and alcohol but such people will continue to be victims of their bad habits even in the absence of compensation money. It is unfair for the rich people to make such blanket statements. A majority of gas victims have used the money in the right way, to make their homes and for their medical care. The amount was so little that most of it got spent in clearing loans or children's health care or in getting married. Even with me the money did not last for more than 2 months.

Illness is removed by its roots

The medical care is pointless because nobody knows the cure for the poisons. When Sambhavna started, only a few *bastis* were given medical care. But later they saw that people in the water contaminated *bastis* are also suffering and they now provide care to many. In a way it sets an example for the Government and shows them that this is the right way of giving treatment and health care. There are proper rules and ways of giving out care. Ayurvedic treatment takes time to give relief but it does give relief and the illness is removed by its roots. Not like English medicines which give you instant but temporary relief. They are also giving health care to the water contaminated areas and this is also one of our demands since the people who are gas victims had to inhale the gas just that one night, but the people living in these *bastis* drink contaminated water all the time every day. Sambhavna is providing them care and the government does not even acknowledge them.

Nawab Khan

I used to have a pain in my knee for a long time and would take medicines. The medication that I got from government hospitals made me very sick until I moved to ayurvedic medication. I came to Sambhavna where the doctors gave me an oil to massage every day for 10-15 days and I am much better and at peace.

CM, protect your sisters from poisons

The people believe that the waste in the factory site is only that which is lying on the ground in packages. The real waste was buried by UCC underground and that is what has to be cleaned. We have seen in other places like Madras where toxic waste was shipped back to America by Unilever. So we need a proper scientific assessment of the situation. When Dow bought UCC it purchased it with its liabilities. However, we have seen that Dow has gradually changed its stand on the clean up issue because of the fight on the ground. First they did not even want to talk about a clean up and now they are ready to clean up but without accepting liability and with conditions to have the case against it withdrawn.

The Government has to provide jobs at the earliest not only for economic stability but also because if the victims have work they will remain active and healthy; they will also be distracted from their troubles. Many victims are falling into depression and resorting to extreme measures; take Sunil for instance, he raised his brothers despite the death of his parents and fought with so much courage but he lost hope and killed himself eventually.

When Babulal Gaur came into power people expected a lot of things but he was the biggest let down because he did more harm to the cause of the victims. Regarding water contamination there was double talk from the Government, the Pollution Control Board found extremely toxic chemicals in the water and the state government refused to accept the presence of toxins. In a unique gesture once when Gaur was the CM the women victims visited him and tied a *Rakhi* (holy thread) on his wrist with a hope to urge the CM to protect his sisters (as it is done traditionally in India) from poisons. When Gaur became the gas relief minister he forgot his promises and the women from his constituency visited him again to remind him of the promises he made as a brother, to provide water from Kolar Dam. But instead of helping the women Gaur registered a case of trespassing against them.

Once we invaded the gas relief department on the issue of clean water and with the demand to start the pipe laying work on the Kolar Dam project, the money for which was already released by the central government. That was probably the most powerful demonstration and work on the project started immediately.

What is the point of growth and development over dead bodies?

The Government has the misconception that the poor only want compensation. It is not true, the poor people want health care, jobs, better employment, participation in the government etc.

The Government has an extremely negative position on the issue, the Chief Minister recently stated that the people in the gas *bastis* are sick because they are living in unhygienic conditions and not because of the gas. The Government is indifferent because it does not feel for the cause of the survivors. This was clear in the Prime Minister's statement to the previous *padyatris* "We need growth and development and things like this can happen in the path of getting there". But has the Government ever wondered what the point is of growth and development that occur over dead bodies? It is true that development is needed but with safety. Not with death and destruction. Development is for progress and prosperity and not destruction. If the latter is the case then one does not see development but destruction. I remember that in 1970 Shakir Ali, the Communist Party MLA was the only one to object to the siting of the Union Carbide factory. Today development in India takes place by compromising the lives of the poor. People in large numbers are falling sick, T.B, cancer etc. Companies open up saying they are providing help and work for the people. But how long does a person work there? 5 to 10 years? And then he falls ill and can't work any more. The company fires him for a healthier worker and it's a vicious circle that goes on and on. There can be development which benefits the common man and our country.

Nawab Khan

Even today the Government is trying to bring foreign companies and no one is trying to stop them. It is only ICJB who is fighting against them. It is helping students to realise and making them aware. And even they have also started joining ICJB and giving their solidarity. ICJB is trying its best to go to schools and colleges to spread awareness about these companies. And also about the Bhopal gas tragedy.

The chemical ministry has recently been caught taking a huge sum as a bribe and registering Dow to sell Dursban. Whereas Dow has been fined *lakhs* of dollars in America, if our government had any shame then they would have immediately cancelled their license for selling it and looked into the matter, not waited for some one to raise their voices against it. But they have not conducted an investigation neither have they cancelled the registration. The Government makes those who don't give bribes follow the rules and those who give bribes are given open rights to violate rules and regulation. If companies want to set up business and really want India to progress then they have to give an undertaking that they will take full responsibility of any accidents that might happen in the future. I bet none of them will be ready.

In India everybody looks up to the courts for justice and it was also true in the case of Bhopal, whatever relief that we got was because of the courts. But the process is extremely slow and time consuming, the victims have spent 23 years fighting and now we wish that the next generation will take this forward. We have gathered this experience over the past 23 years but the new generation will be able to grasp this in a matter of a few years and a lot of young people are coming forward to join the struggle. The new generation will keep an eye on the Government and the multinationals.

The day 2 *lakh* victims take to the streets there will be victory

For many years the fight of the victims was only limited to within Bhopal and now we have managed to get so much international support and I am very proud of it. If the people of Bhopal learn to utilise all the help they are getting from around the world to their advantage then the fight for justice will succeed. The victims are unable to participate because they are unemployed and weak and they have lost hope because so much time has passed. They have lost the faith in victory. This is wrong, there are 5 *lakh* 75,000 gas victims and the day 2 *lakh* take to the streets the Government will have to listen.

The politicians are spreading rumours about disputes between groups but I would say that there are absolutely no disputes between groups. They are all working in their own way and they also meet up regularly. Disputes occur only among political parties and politicians. Nobody leaves the fight because of differences. Some take a break, some get sick or old and some change groups, nobody quits. People also ask me why I continue to fight. I tell them that this is my habit, some smoke, others drink but I fight.

A non gas victim would only marry a gas victim if her family is rich

I was born in 1947 in Sironj, Vidisha District. I spent hardly 5 years in Sironj, my father was in the Border Police and was transferred to Rajasthan; I went with him to pursue my education. In Rajasthan I studied English, Hindi, Arabic and Farsi and I also passed exams in Aligarh Muslim University. I moved to Bhopal in 1960 to learn tailoring and then I stayed back. I met Maulana Aqeel, the general secretary of Jamat Ul Hind in Kota who encouraged me to write in newspapers and magazines at the age of 17.

My family was into farming and was pretty well off. My brothers and sisters lived in the village where we had 250 *bhigas* (a traditional unit of measurement) of land and 4 -5 vehicles. My uncle was a farmer and my dad worked for the government liquor shops. I came to Bhopal because I have always liked travelling and since all the other responsibilities were taken care of by my brothers I did not feel happy in the village. I was married in 1970 in Bhopal because my wife was from Bhopal. I went back to my village for a year and came back in 1971. The oldest son was born in the village and my six daughters were born here. The oldest son died, and I got 4 of my daughters married. I was forced to find gas affected men for them because we were poor and gas exposed. A non gas victim would only marry a gas victim if her family is rich and economically stable, a poor gas victim will have to settle for a gas victim for a partner.

One might have thought that family pressure on the women who participate would have held the movement back from taking on bold or dangerous actions but when women took the lead they shook companies and governments. During the *Jhadoo Maro Dow Ko* (beat Dow with brooms) women attacked the Dow office in Bombay and they also demonstrated at the high security Bombay International Airport. Nobody had attempted that but women from Bhopal have done it.

My participation in the campaign has not been affected because of my religion. Such divisions only happen among the rich Hindus and Muslims. In our religion the noblest deed after *Namaz* is to serve other people irrespective of their religion: "*Khidmat e Khalq*", and in the *Ramayana* there is a quotation "*shantakaaram bhujaga shayanam padmanabham suresham*" which means that any deed done by a person of any religion with "peace" in mind will be blessed with success by God". Similarly in Farsi a quotation which means "any person who pays attention to things will succeed in life" – I would use this is in relation to the movement. The *Quran* clearly says *Khidmat e Khalq*. *Khalq* means 'the world' so then how can I say that I will not serve a Hindu? God says *Rab D Ul Alameen* which means I am God of all the world and not *Rab D Ul Musalmeen*. In our organisation harmony is the first lesson and we are all from one big sad gas *peedit* family.

At the time of the riots following the Babri Masjid episode I was residing in Gondipura where there is a mixed population of Hindus and Muslims and we managed to maintain peace there. No police were allowed into the area because we did not want trouble and we could look after ourselves. I feel that the riots have brought people closer to each other.

There are people in this country who support justice

We do not organise programs without consulting the public. All the ideas for programs come from the people, we do not believe in taking charge because we are organisation leaders.

Dow had plans to start a research centre in Bombay and wanted to recruit students from a university (IIT). We went and spoke to the students about the plight of Bhopalis and also the attitude of Dow and they all decided to boycott the company. So I feel that there are people in this country who support justice and I also do not blame the people who oppose our cause and support Dow because there will always be a small percentage of people on the bad side.

The old people are in the forefront of the fight now because they are more experienced and they have also had a considerable contribution to the fight. A lot of young men and women are coming forward and they should start a new group, we will guide them with our experience. We have also hired a young person for school outreach in India so that young children learn about Bhopal.

I have not faced any problem or difficulty in my work because when I get sad about the loss of my wife and son I always come across people who have lost many more loved ones than I have. This makes me forget my sorrows.

We will not be disappointed if we do not get victory, I believe that despite the delay we will certainly get justice. It will be a blot on the history if Bhopal fails to get justice, and our country, law and system will be shamed. The fight of Bhopal will get tougher in the future. This is no longer a fight for the gas victims but for India, everywhere that injustice prevails. This has become an international movement and we foresee a long fight and our younger generations will continue to fight.

S.T.D.P.C.O
ANAND PUBLICITY
S.T.D P.C.O
मेडीकल जांच रिपोर्ट
भोपाल गैस पीड़िता का
मुआवजा

Shahid Nur

Bhopal Ki Awaz
Voice of Bhopal

Shahid Nur was orphaned by the gas disaster. As a young adult he formed a campaign group of fellow orphans *Bhopal Ki Awaz* in which he took a high profile leadership role. He has since became less active on the Bhopal issues as he focused his attention on his own family's needs.

The tragedy took our parents away

It was a Sunday so my father was at home. We lived opposite to where the *aloo* factory is now located, right beside the factory. We did not have a TV so we used to go to our neighbour's to watch television. We returned from there and read the *Hadees* and went to bed. Then at around midnight something weird was in the air and I shouted out. I ran out of the house and reached an open area in the front of our house into the gas clouds. My eyes began to burn and I ran towards a tank of open water and rinsed my face which gave me some relief. I was around 9 years old then.

We were separated from our family and I with my younger brother and sister headed to the station. At the station my brother got separated.

My mother was trying to board a truck heading out of the city and was killed when the truck began moving while she was still trying to get on.

My sister was raised by my mother's mother and now works at Vardhaman Fabrics in Budhni. My brother and I were raised by our father's mother. My grandmother had a flour mill at home so we ran that and also went to school. Our family was sustained on the flour mill and some donations that came from outside. We got free food from the Government. I studied till the 11th grade then quit school in 1993. I quit because I could not afford the fees of the private school. My brother now runs a cell phone shop in Bhopal.

The biggest difference the tragedy made was that it took our parents away. Both my parents were educated but we could not get good education, only my sister completed her graduation. The kind of things we did after our parents died we never did while they were alive. When they were alive we had no work to do, all we did was go to school, play, eat and sleep. Then all that changed – we would wake up and groom the buffaloes, go to school, then manage the flour mill. We would have done better if our parents had been alive.

My father came from Bhopal. He was an educated man, he had done his BA, after which he was employed as a watchman with the Food Corporation of India. My father got the contract to maintain the grass fields inside Union Carbide and the factory behind Carbide called Shama Forge. He had people working under him and we received all the grass from the factory. We were also invited for all functions at the Carbide factory.

I had heard that a lot of workers from J P Nagar got sick after working in the factory but I have no confirmed reports. My father's friend Kashi Ram worked as an electrician inside the factory. I learnt a lot of things about the disaster much later after I heard people talk about it.

My wife's parents also died in the gas leak, she is an orphan. The Government took us on excursions and we met at one of these trips and then got married.

I have three children now and they do not go to school. Education is very important. It's better not to be educated than go to government schools. The children go there just because they get free food; the quality of education is abysmal.

In the early days everyone participated in rallies

I did not even realise when I joined the movement. The biggest demonstration was outside the Chief Minister's house when all the children participated and we used to live there outside the CM's residence. The other organisations had done this and I also reached there with a lot of children from my *basti*. This was right after the gas leak and I don't remember the CM, I guess it was either Motilal Vora or Arjun Singh.

During the first anniversary when it was all fresh in our minds half of Bhopal participated in the rally. I joined that rally and then my participation kept growing. I was away for a while in between and then I joined again with more determination. I don't remember how and when I joined as a youngster. My friend Sunil started an organisation called Children Against Carbide which I was a member of and used to join the demonstrations. Sunil was about 13 or 14 years old when he started this and I was younger than him. As part of the movement, Sunil took part in everything that happened and worked with the other organisations. It was Sunil who inspired me and because of him I joined the fight. And now he is no more.

I was usually active during the anniversary day rallies and later I also began participating in rallies to Delhi. I initially participated in a lot of rallies to Delhi organised by BGPMUS. I would go with my grandmother and my aunt who were part of the movement.

Initially the demands were only compensation and medical care. Later the demands of clean up, clean water, economic rehabilitation and pension were included as the movement progressed. I don't remember the dates but I can give you the sequence: first economic rehabilitation was raised; then clean water; then clean-up of waste; then very recently was Dow to take responsibility.

In the early days everyone participated in rallies. Children would see a rally and just join in: that is how I started. Nowadays, people just watch, they hesitate to participate. This is because the issue is getting old and people are forgetting about it and I feel that people have not learnt what they were supposed to from the disaster. The fight for the second compensation attracted a lot of people, huge rallies went to Delhi. When the compensation was disbursed people slowly trickled out. Participation is very unpredictable, if people feel that certain issues can be won with more participation, they join. Similarly the 25th or the 30th Anniversary will see huge participation

There are many organisations fighting on different issues and they have kept going because the demands are not being met, even if they are promised then their implementation remains an issue. Like the demand of employment, the Prime Minister's Office promised employment but there has been no implementation. Similarly tenders and orders for clean water from Kolar dam were issued but then it was blocked in the implementation stage.

Other smaller issues like pension also remain undone. It takes years to get a pension book made, lower level government employees make mistakes with names and reject the forms thus harassing applicants.

The youth are not involved in the fight but it is very important to get their fresh perspective into the struggle. It motivates the older generation and also brings in new ways of fighting.

So people will join only if they see that work on these issues is done. Nothing is moving on these issues and people just continue to participate in rallies and are now tired. This does not mean that people should give up fighting; all that has happened so far has been because of the fight. Now at least the gas victims get some audience for their problems. If the fight stops they will completely stop catering to them.

It has been 25 years mainly because there is a lack of political will. The main demand of compensation was partially won through the court case although the demands of the movement have not been met so far.

People have given up after losing patience; new people have joined over new problems like water and second generation victims. It's a process, problems will not end soon.

Gas affected orphans' movement

I don't remember when or how I joined but there was this organisation named Bharthi started by Vicky Lamba, a journalist and a social activist who was associated with the Congress Party. He used to take us for meetings with the governor or the CM for employment. He wanted to help us and he felt that the best way was to get our education sponsored. Initially a lot of help came in-kind in the form of pens and pencils etc. He gave the responsibility to educate us to some people in Bhopal. A few children were sponsored by individuals for a couple of years and this eventually stopped.

All the Government did during this was to invite us for annual functions where the CM or the Women and Children Development Minister were present and they would display us and give us some clothes. The papers next day would have news of Government aid to 28 gas orphans. This happened every year, the annual report would have the details of the budget for instance saying Rs.400,000 was spent and the surplus was returned.

The scheme supported about 28 children (who received goodies once a year) and four children who had no relatives were given full support. They lived in a hostel called Kalyani, all their needs like food, clothes and education was taken care of and they also had a governess to look after them. I feel that this was a sort of discrimination. These four children had no one to look after them so the Government took full responsibility. We had our grandparents who did not want us to be raised as unsupported orphans so we got the least amount of help.

Mr. Lamba quit in 1994 because I guess he got too busy with Congress or maybe he stopped after he got an award and thought he had done what he had to and received an honour for it. I still meet him, he lives across Apsara Talkies in Bhopal. When we were with Lamba's organisation we started to organise and then around 1993 we separated from him and started our own organisation named *Gas Peedit Anath Bachche* (gas affected orphaned children). More than half the people who were under that scheme were with the orphan's movement: Sanjay, Suman many others. We were not very active at first in terms of demonstrations but we met quite often. There was no deliberate motive for the organisation's name, we just started calling ourselves 'orphaned children' and started using that name on the banner. We used that name until 2004.

Five of us from *Gas Peedit Anath Bachche* went for the World Social Forum in Mumbai in 2004 along with around 200 people from other Bhopal organisations. There we met students from a Delhi college and they suggested that we name our organisation *Bhopal Ki Awaaz* (Voice of Bhopal). We liked that and we renamed the group.

We used to meet people from all the organisations, we worked with Alok Pratap Singh of Zehreeli and he used to bring people to interview us. We worked with Sathyu, we worked with David Bergman, he gave us comics to read and taught us how to play the flute!

We have done a few small demonstrations because our group is only 28 people strong. Chief Minister Digvijay Singh promised employment to victims who passed their standard 10 in school but nothing happened, so in 2003 we held a seven day hunger strike at the tin shed over the issue of employment. As a result the CM met us and said that he cannot promise us jobs but he will give us an *anudaan* (grant) to start a business.

Similarly when Babulal Gaur was in the opposition he promised us jobs and he also helped us get our confiscated tent back from the police but when he came to power he found our demands unreasonable.

Both the political parties are in on the side of UCC because what happened did happen because of the political parties. Arjun Singh from Congress was the CM when the disaster happened and Babulal Gaur was a lawyer for UCC at that time and he is with the BJP now. These are the only two major parties in Madhya Pradesh and both favour

UCC. Congress spokesperson Abishek Manu Singhvi is the lawyer for Dow in the clean up case.

All the governments both state and central have done nothing. So far there has been no special scheme for gas victims whether pension for widows or jobs for youngsters or exclusive medical facilities for victims. However I do vote and I do not feel that boycotting elections makes a big difference because when we did it once it just reduced the overall percentage of votes. Only if the whole of Bhopal boycotts elections will it make a difference.

The Government has no right to criticise the organisations and their politics. They should first fulfil their responsibilities of clean up, water, pension, medical care etc. There would have been no need to make so many organisations or to raise funds or provide separate medical care if only the Government had fulfilled its responsibilities.

I never felt that there has been discrimination in the treatment of survivors based on religion, but I do feel that the BJP is demanding the additional 20 wards to be included in the gas category because it is a Hindu dominated area. It is vote politics. A little discrimination crept in after the riots (in 1992, after the destruction of the Babri Masjid in Ayodhya) but in old Bhopal there is not much. One cannot say everything is perfect because the riots did change the dynamics but it's not as bad as other riot-hit places.

Differences in the movement are not huge

I feel that the organisations split because of differences mostly among the workers and rank and file, the leaders have no major differences. All the organisation leaders are part of the National Coordination Committee and are working together and are united in their demands. When the demands are presented to the public they seem different because they all have different priorities so people think there are disputes. Despite the united stand during local demonstrations the groups prefer to remain separate because there could be fights between the memberships. Their differences are not huge and if need be they will all come together: they have in the past but they separate over

Shahid Nur

small things like the placement of names on the banner: this is a major reason for disputes within coalitions who fight to get their names on the top or in the front in a rally!

Some groups will take international funds. There is nothing wrong with international support, if people want to pitch in they should be allowed to in whatever way they can. There are other organisations who believe in raising funds locally from gas victims and within India and yet others who take assistance from the Government. They all have their own unique ways and no way is wrong as long as it is spent for the victims. None of them has a lot of money to spend so they have to raise funds somehow.

I have been part of all major ICJB campaigns and I must have participated in about 50 demonstrations so far. I have also worked with Jabbar *bhai*. Each organisation has a different way of working so a person has to make some compromises if work needs to be done and one needs to get along with any organisation. Some organisations have their leader take all decisions and others consult everybody democratically. I feel that the right decisions come when more people are involved, but it is also possible that a decision taken by a single person is right.

The demand for clean water was first raised by Bhopal Group for Information and Action. The fight of gas and water victims is against the same company so it is the same. The other organisations have also included that now. The problem is that the organisations do not have the resources to get the water tested and the Government lies about it.

Not all the organisations want to see Dow clean up the site. *Zehreeli Gas Kand Sangharsh Morcha* has filed a legal petition to demand a clean up as soon as possible, even if it is paid for out of government money. However, going by the principle of 'polluter pays' Dow has to clean up, and hopefully there will be a law on an international level that will insist on this. If the Government wants to clean up it could have done it back then in the beginning. It wants to do it now to let

Dow off the hook, so that Dow will not have to take responsibility for the damage caused by the waste.

If you or I purchase a product we acquire its good and bad aspects. Similarly when Dow acquired UCC it also acquired the liability of Bhopal. Recently we protested against a business deal between Indian Oil Corporation (IOC) and Dow in which Dow was planning to sell Union Carbide's technology to IOC (the deal was called off). So Dow finds nothing wrong with making money from UCC but it does not wish to address the pending liabilities of UCC.

The Supreme Court decision on compensation after the settlement clearly said that the government must make up the deficit in the compensation figure. Total compensation was calculated on the initial figure of 1 *lakh* 5,000 victims and it was distributed among 5 *lakh* 72,000 victims, so people hardly got anything. This is the basis for the '5 times more' compensation claim made by Jabbar *bhai* and BGPMUS.

Sathyu knows the most about the movement but he does not meet the public very often. Rachna is good in the field. Jabbar is also good with the campaign information. They all work for the people, no one is doing anything wrong. If I were to choose an ideal campaign leader I would want Sathyu's understanding of issues, Rachna's field work and Rasheeda Appa's oratory skills.

Dow is America's favourite company

For me, victory for the movement would be a clean up of the site, employment for the survivors, clean water and a pension enough to run a household: around Rs.1,000 to Rs.1,500. Victims who are bed-ridden should be provided with medical care at home by the Government. The fight of Bhopal is not just about compensation it is first about delivering justice by punishing those responsible for the disaster. Looking at the situation now even if Anderson is brought back what kind of punishment can be awarded to him, he is so old. I feel that even if Bhopal gets justice it will be incomplete. Dow is America's favourite company, it will never be prosecuted.

Shahid Nur

I do not think that a collective of even 10,000 or 15,000 survivors will be able to make a big difference to a coalition of powers as big as Dow, the US Government and the Indian Government. All the influential people have compromised with the other side.

Sarita Malviya

ICJB Children against Dow Carbide

Sarita Malviya was interviewed in 2007 aged 14. She has become a leading advocate of youth activism, especially amongst those who continue to be affected by contaminated water. As this book was being compiled, she was touring in the USA to raise awareness amongst American young people. Intelligent, articulate and angry, Sarita is a likely candidate for the next generation of leaders.

Once someone is a part of a cause then the spirit to fight comes on its own

I was born in my grandmother's village. My parents are day-labourers and moved around in Bhopal when I was young until we settled in Prem Nagar. My health is affected by the pollution in the water from the factory and I take ayurvedic medicines prescribed from *Sambhavna*.

I started going to the weekly public meetings at the *aloo* factory with my mother, and then started getting involved the campaign. Now I attend every week. I sometimes bring some friends with me. Occasionally I forget and go to school instead, and then my mother updates me. I have become the main link from the meetings to my colony. I make a point of telling others what is happening and encourage people to participate in the rallies.

When I'm on my way back from the meeting, some people ask me about what has been going on, and soon a crowd gathers and so I get to inform many people. When I go to collect water the women ask me. The men ask me as and when I meet them. There are some women and men I know who are interested but can't come so I go to them and tell them.

I tell people what I have heard in the meeting, how I understand the issues and other discussions which have gone on. I share all the information from the *aloo* factory and also by asking Rachna or Sathyu when I meet them or I come in to *Sambhavna* to ask them. During the *padyatra* I would keep my neighbours up to date by finding out details from the *aloo* factory meeting.

In the public meeting I also tell stories about what is happening in the colony and any problems that are happening. I'm always listened to and my opinion is taken seriously when I give it.

My generation is the future of the campaign

Although we did not live through the disaster, my generation is the future of the campaign. The campaign needs to draw on the help and support of the youth because the government and Dow need to see that the campaign is not limited to the older generation. The new generation need to be educated about the disaster and the need for justice. The boys who hang around the rallies and mock the older activists, they don't know about the disaster. They would also join if it were explained to them. I try to explain to my classmates about getting involved. That same spirit to fight can be there amongst the youth if they are made aware of what happened and of the difficulties families face even today. And once someone is a part of a cause wholeheartedly then the spirit and the strength to fight come on its own.

My parents are day-labourers. They didn't have the chance of education themselves, but they value education for me and my brother. There are many other children who are poor like us who don't go to school because they have to work. My family is very committed to education. I want to be lawyer and to use the law to fight for justice.

The teaching is all right in our schools, but it is just that the government schools have very little in the way of facilities. Some schools have them and some don't, and at times teachers are also missing. Whatever I am taught I think it will help me in the future. The school supports me in my involvement in the campaign. When I miss school to go to meetings or rallies I always make up lost lessons. Teachers don't treat me any differently because of my involvement in the campaign.

Men are not the only people in this world

It is particularly important for a girl to be educated equally and made aware of her surroundings so that she knows her rights and cannot be pushed around. Men are not the only people in this world, women are equally important and in many ways we are ahead of them. Women can go ahead and study, work or fight.

I know about other campaigns in India mostly through the work of women, for example in Orissa and Nandigram. Solidarity with them is very valuable and I would fight alongside others where they have been denied justice.

I think the next step for our campaign should be to block Dow's growth in our country from all sides: surround it, so that it has nowhere else to go. And also none of us should be voting because it just gives credibility to the Chief Minister whoever it is. They are all the same. There is no point bringing to power a politician who comes and speaks to us, makes promises and vanishes once he forms the Government

Governments deny justice because they are in the pocket of the multinational corporations

I'm not against government but against their lack of justice. I will support any government which gives justice. I would like to throw a slipper at the Gas Relief Minister – why doesn't he get it? He blames our ill health on dirty living, but he should come and live in our *basti* and drink our water. I get so frustrated.

Governments deny justice because they are in the pocket of the multinational corporations. MNCs and other foreign companies shouldn't be allowed to come to India. If they do they should be obliged to care for people and the law should be implemented. All companies and their scientists should be responsible for their inventions. Poisons should not be made, or if they must then they should make less and make an antidote. It is possible to live without chemicals. We should stop buying chemicals.

It's not just that the companies are owned by foreigners. I know that the campaign and Sambhvana Trust run on money donated mostly by people from outside India and this is completely ok. They can earn a sufficient amount and still take out some money and give to us, who cannot even get one decent meal a day. The money that is given to us is given by choice and we don't demand it. In this world every one relies on something or some one so it should not be seen as a problem if we do the same. But we should not rely on this money all the time and look for other means, one of which is the Government. It is their responsibility. The government should take notice and help us and give us our rights.

Many foreigners come here to make books and films to tell our story all over the world and that is very important. Even if they make a little money that's ok. Young people depend on these to learn about the leak, and it is good to have solidarity from these people.

The young are the ones to take the movement forward

I have faith in God, although it is true that many who have suffered don't. Many people died and are suffering and hence their faith is shaken. I have not suffered in the same way and I do believe in God. When Rachna and some others were arrested when they were on hunger strike, I did pray for their safety and quick release, and also that they are not forced to eat. And then too I was heard.

I have seen people from the church coming and helping but not from temples or mosques. I think the religious institutions and parties should support us more but making sure that they do not bring any kind of differences amongst us. Any political party which decides to support us can do so the way we are, with no differences. Any body who wants to support but split us we don't want their support. Communalism has never been an issue in the campaign, nor caste divisions, in either older or younger generations.

We will keep fighting and justice will come eventually. Never lose hope. Defeats only motivate us to fight harder. I am inspired by people who have been fighting for 25 years. Now the young are the ones to take the movement forward.

Sarita Malviya

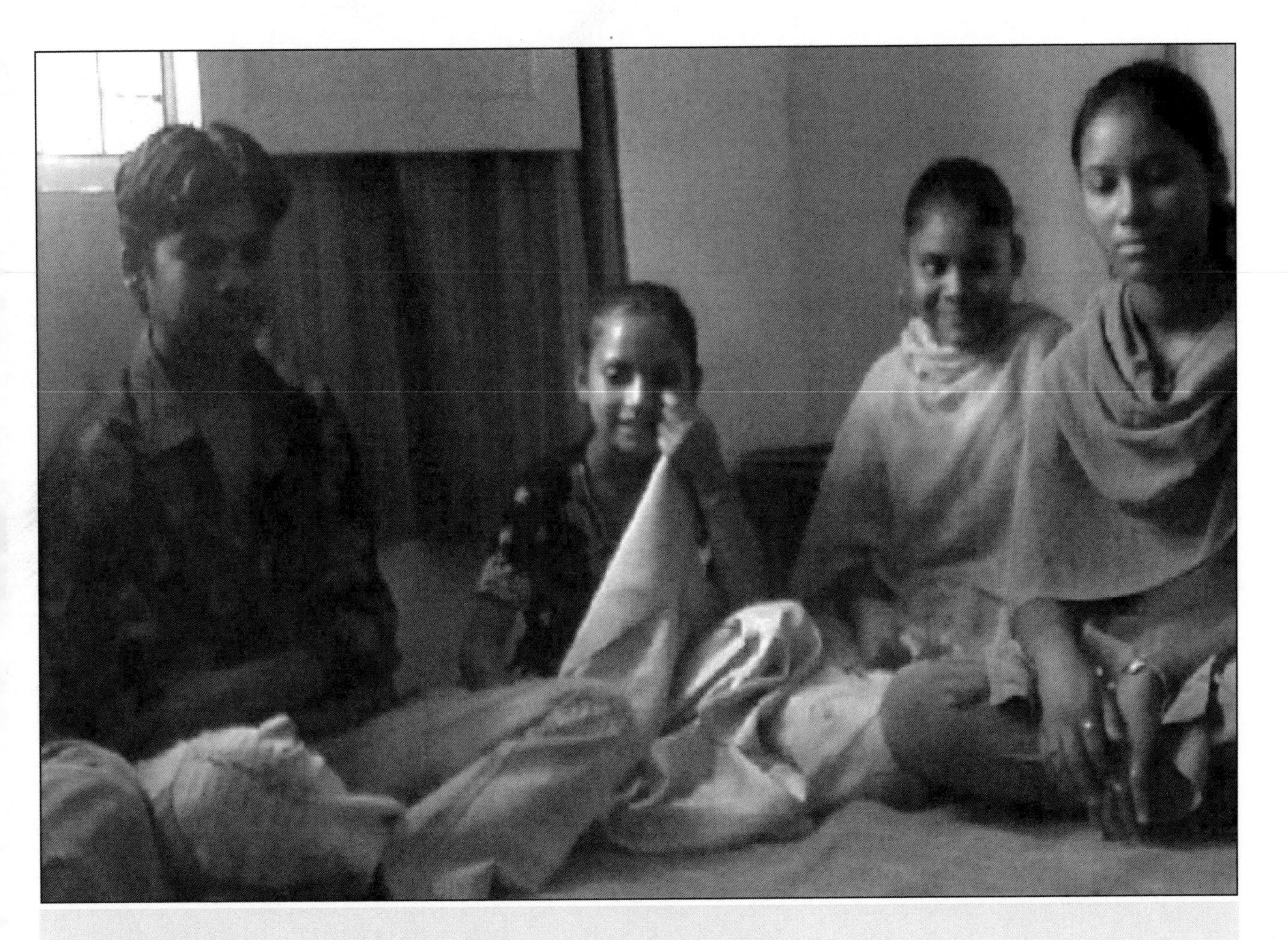

Children against Dow Carbide

Komal (17) Amir (15) Nahisha (7) Yasmeen (12) Kiran (15) Kajal (9) Pooja (12) Reshana (12)

Whilst the research for this book was being completed, a group of children and young people formed an organisation to have a distinctive voice in support of the campaign. Attracting up to 50 to meetings at times, a small group agreed to be interviewed some days after their first major demonstration. Represented as a group interview, the voices are mainly those of Amir, aged 15 and Yasmeen aged 12, passionate and articulate voices who demonstrate that the future leadership of the movement is emerging..

We organised the first rally of Children against Dow Carbide

We wanted to organise a separate organisation for children within ICJB because we felt it was important for children to be aware of what happened and what is still happening. Also we know that our elders have been fighting for 25 years and have had no results but we have seen that the Government pays attention when children are involved. We tell other children whenever we can about the disaster and how we don't have justice. Some of us took part in the *padyatra* (from Bhopal to Delhi in February 2008) and that was when the idea came up to start our own organisation.

We have elected four children as leaders. They were selected by a general consensus because they know most about the issues. Names were put up on a blackboard and everyone agreed that these would be our leaders. Their role is to make sure that all children are aware of what is happening by making announcements, and to make sure that nobody misbehaves. We spread news of what we do by word of mouth in the different *bastis* and sometimes we go to the houses of children we know who can pass things on.

We organised our first rally of Children against Dow Carbide which took place last week (May 9th 2009). Two of our members have gone to America to try to speak to leaders of the company there and there was a rally there too. We selected those two by an election. Those who can speak well and their parents would allow them to go, they gave speeches to everyone and they were selected on how good they were at speaking and how well they were able to explain the issues.

So many children are disabled as a result of the gas tragedy

The children's *sangathan* is distinctive from the other groups although we are all united for the same cause. We get help from the other groups: for the rally we got help with having the banners printed, deciding the route of the march, contacting the media and so on.

Children Against
Dow Carbicide

We meet every Sunday and the children decide what it is they want to do. We write down on a chit what we want to do and then decide from what is the most popular thing. We always start with a discussion of our aims and news of rallies and things. Then we do what we have decided to do: watch a film, go on a visit, share some food etc.

Our teachers at school know what we are doing and are very supportive. The principal of one school is particularly supportive: when we went for the *padyatra* he helped a lot by postponing exams and in other ways. Other children are very curious about the *sangathan* and ask us why we are doing this. We say it is because so many children are born disabled or damaged, with fingers missing or not able to walk, as a result of the gas tragedy. This is very sad and we want them to have justice. The other children are moved by this and ask 'Is there anything I can do to help?'

Even when we get justice we will keep fighting

We have the same aims as the other organisations and have become aware of the issues from them. We are particularly concerned about the disabled children and want to make sure that they get their rights to compensation and health care. We are also concerned about children affected by the poisoned water. We would also like to campaign for those children who are not able to go to school because they are so poor and need to work, but it would depend whether they wanted to go to school. We would ask them first.

Companies from outside India shouldn't come to India. We should just have Indian companies here. If the factory had been owned by the Indian Government then we would have got justice by now. We also think that Dow Carbide should clean up the factory site. They should take all the polluted waste back to their own country. If they had had disabled children the way we have, then they would feel bad and get justice. If the Government cleans up the site then the company will feel that they have got away with it and some other country will become the next Bhopal. Big companies are important for India to progress, it is true, but they should not build factories in populated areas. Not all companies are bad. Some make tractors or things that are good for development.

What has happened has happened and we can't change that but we will keep on fighting. We want to stop another Bhopal happening elsewhere. Even when we get justice we will keep fighting so that no company feels it can do what Union Carbide did, and nobody else will have to experience what Bhopal has.

Style

JUSTCE IN BHOPAL
MEANS
ASAFERWORLD
FORALL